AF765600

Institutions and Incentives in Regulatory Science

Institutions and Incentives in Regulatory Science

Edited by Jason Scott Johnston

LEXINGTON BOOKS
Lanham • Boulder • New York • Toronto • Plymouth, UK

First paperback edition 2014

British Library Cataloguing in Publication Information Available

Library of Congress Cataloging-in-Publication Data
The hardback edition of this book was previously cataloged by the Library of Congress as follows:

Institutions and incentives in regulatory science / edited by Jason Scott Johnston.
p. cm.
Includes bibliographical references and index.
1. Environmental policy—United States. 2. Environmental agencies—United States. 3. Environmental protection—United States—Standards. 4. Environmental law—United States. 5. Environmental sciences—Political aspects—United States. 6. Science and state—United States. I. Johnston, Jason Scott.
GE180.I57 2012
363.7'05610973—dc23
2012003620

ISBN 978-0-7391-6946-9 (cloth : alk. paper)
ISBN 978-0-7391-9067-8 (pbk. : alk. paper)
ISBN 978-0-7391-6947-6 (electronic)

Contents

ONE

Introduction

Jason Scott Johnston

The vision of rational, science-based environmental, health, and safety policy: that is the objective of the modern regulatory state that is in crisis. For decades now, there have been charges on the one hand that high level regulatory officials interfere with and politicize policy choices that should be based on science alone (see, e.g., Wagner and Steinzor 2006) and on the other hand that both regulators in justifying regulation and industry in opposing it too often rely upon "junk science." Modern regulatory regimes continue to insist that they are based upon science even as regulatory science—the science that underlies actual regulatory decisions—has lost credibility. The lack of confidence in regulatory science has, with considerable justification, translated into a lack of confidence in modern environmental, health, and safety regulation.

In areas of regulation ranging from climate change to chemical safety and the protection of endangered species, the chapters that comprise this book provide detailed case studies of the crisis in regulatory science. In so doing, they systematically identify certain recurring structural features of existing regulatory institutions that have created very fundamental incentive problems. Current laws and institutions for regulatory science have made science so significant in regulatory policy that they have created both an incentive for regulators to cloak policy decisions in the mantle of science, and have also—by failing to understand the uncertainty inherent in most scientific findings—put scientists in the position where their own policy preferences guide important choices of scientific methodology and interpretation. Moreover, the contributions to this volume go further by suggesting potential institutional reforms that promise to restore the credibility of regulatory science and to improve contemporary environmental, health, and safety regulation. In this introduction, I briefly describe the chapters while summarizing what I believe to be their most important, common general themes.

THE NATURE OF THE PROBLEM: SCIENCE ADVOCACY VERSUS THE SCIENCE CHARADE

In the law review literature, a very prominent critique of science in the regulatory process made by Wagner (1995) is that the current regulatory structure creates an incentive for regulators such as the federal Environmental Protection Agency (EPA) to hide policy decisions behind science. In the present volume, an example of such a use of science is given by Wyman. In her chapter, Wyman describes the decision by the United States Fish and Wildlife Service (USFWS) to remove the Northern Rockies gray wolf from the list of endangered species under the U.S. Endangered Species Act (ESA). Under the ESA, endangered and threatened species are afforded expansive legal protections that effectively regulate land development and a wide range of federal activities in affected habitats. For organisms to be protected under the ESA, they must be classified as either a species, subspecies, or distinct population segment. As Wyman explains, the actual listing or delisting decision under the ESA involves not only the threshold question of whether the group of organisms is a "species" within the meaning of the ESA, but also a determination of whether that group is at risk of extinction. As she argues, however, estimates of extinction risk are generally subject to uncertainties—estimates may be wrong. Hence the decision whether to list or delist always involves a weighing of the potential costs of the two types of errors—a false negative, a failure to list or retain listing for a species that ends up being at a high risk of extinction, and a false positive, a listing or continued listing of a species that ends up not being at risk. The Northern Rockies gray wolf had been protected under the ESA as a distinct population segment. In Wyman's view, the USFWS's decision to no longer protect the Northern Rockies gray wolf in fact involved a policy tradeoff in which that agency weighed the potential cost of removing the ESA's protection—a possible decline in the wolf's population—against the cost of retaining it—economic loss and attendant political cost in Idaho and Montana, where ranchers believed that the continued listing of the wolf under the ESA was making it legally risky for them to take actions to protect their sheep and cattle herds from wolf predation. In its delisting, however, the USFWS did not reveal the policy tradeoff that drove its delisting decisions, but attempted instead to portray the delisting as justified entirely by scientific evidence that the Northern Rockies gray wolf population exceeded the minimum viable population size.

This is an example of the science charade in action. But the science charade is a well-recognized problem, and while some of the chapters in this volume (such as Dudley and Gray, and Wyman) do discuss the science charade problem in detail, many of the chapters stress a much less discussed but perhaps even more serious problem with contemporary regulatory science: the incentive for regulatory agencies or regulatory assessment institutions to engage in advocacy science. Advocacy science is the practice of culling a complex body of scientific literature for studies or funding new studies that support the decision to regulate, while ignoring or downplaying those that undermine the case for regulation. While the incentives for regulated industry to engage in such science advocacy has been extensively documented (and much decried) by some legal scholars (see, for example, McGarity and Wagner 2008), as the chapters in this volume show, regulatory agencies and other scientific assessment institutions are just as capable of engaging in and rewarding such advocacy tactics.

In their chapters, both Johnston and McKitrick argue that in the area of climate change regulation, the Intergovernmental Panel on Climate Change (IPCC) Assessment Report process has produced not the balanced assessments of climate science that have been advertised, but briefs for one side in ongoing disputes at the frontier of climate science. Johnston cites a number of examples—such as projections of future species loss and the role of black carbon (or ordinary soot) in Arctic warming—where IPCC Assessment Reports present a picture of the current state of the science that seems drastically at odds with what one finds in a survey of the peer-edited literature. Drawing both on the Climategate emails and on his own experience serving as an outsider reviewer on the temperature measurement chapter of the 2007 IPCC Assessment Report, McKitrick clearly shows how the IPCC process enables Lead Chapter authors to actively advocate their own views in IPCC Reports and dismiss or ignore contradictory views. For example, a draft of the chapter on temperature measurements that McKitrick commented on included cautionary remarks on the dangers of inferring statistical significance from time series trends where it is known that random fluctuations in temperatures tend to persist for several years. However, the lead chapter authors simply deleted those cautionary remarks from the final version of the Report; such behavior was hardly unexpected by the IPCC, however, for one of the Lead Authors for that chapter, Phil Jones of the University of East Anglia, produced the temperature time series data that was under attack in the peer-reviewed literature.

In the area of species regulation, Ramey's chapter explains how there are now a multiplicity of scientific standards that now coexist within biology for the definition of a "species" and "subspecies," the assemblages of organisms that are protected under the ESA. He recounts several very high profile disputes over listing under the ESA—such as the listing of the Prebbles jumping mouse in Colorado and Wyoming, and the listing of the coastal California gnatcatcher—where there were competing scientific studies in the peer-reviewed literature on whether or not the populations at issue constituted a subspecies. As Ramey shows, the USFWS—the agency which administers the ESA—has seemingly tilted the scales in favor of studies and methodologies that favor finding a subspecies, triggering ESA listing.

In the area of environmental regulation, Jamie Conrad's chapter, beginning the third part of the book, presents a number of case studies focusing on chemical risk assessment practices at EPA and the National Institute of Environmental Health. His case studies clearly demonstrate these agencies engaging in the tactics of science advocacy. EPA's risk assessments (or in some cases reassessments) for formaldahyde, dioxin, and inorganic arsenic, among others, all reveal a pattern in which the agency continues to rely on studies that support its conclusions that the substance is potentially carcinogenic even at very low doses, even when a large body of much more recent work has shown that the assumptions underlying EPA's preferred studies are likely wrong and their results unreliable. When new research is funded, as in the case of the NIEH's funding of research on bisphenol A (BPA), the agency is clearly attempting to get studies which support its preferred result—serious adverse health effects even at low doses.

THE INSTITUTIONAL INCENTIVES FOR REGULATORY SCIENCE ADVOCACY

The summary I have just provided gives only a quick glimpse of the fascinating case studies of regulatory science advocacy presented in the chapters below. These chapters also constitute the beginning of what I hope will be a sustained and continuing program of research into the institutional causes of regulatory science advocacy. The search for institutional causes is really an investigation of the fundamental issue of how alternative *institutions* for regulatory science impact the *incentives* of both regulators and regulated entities to engage in advocacy science. Virtually all of the chapters in this book address this fundamental issue. Dudley and Gray believe that the current regulatory institutional environment is one that insulates agency experts from dissenting scientific views, reinforces prior beliefs and biases regarding the relevant science, and produces regulatory decisions that lack transparency. In their view, the design of institutions for regulatory science, or what they call "science policy," involves questions of which science to consider, how to evaluate multiple studies, and what to do in the face of the uncertainty generated when there are competing scientific models and evidence. Dudley and Gray argue that when one puts the typical federal regulator's lack of knowledge about science and the scientific method together with such a typical environmental or safety regulator's general incentive to err on the side of overregulation (both to avoid potential criticism for doing too little, and to build a reputation with interest groups that leads to future career opportunities), one gets a regulator who will find it extremely attractive to justify regulation as based on the "best science available." As Conrad also argues, such an agency—naturally enough staffed by people who believe in the agency's stated protective mission and who for reasons of institutional and personal self-interest tend to strongly prefer to find substances to regulate—will also have an incentive to cherry-pick the literature for studies that support risk estimates that are large relative to those in the literature as a whole, and then present those estimates without adequately explaining the many reasons why they are uncertain. Thus both the Dudley and Gray and Conrad chapters believe that fundamental agency objectives account for the incentive to engage in both the science charade and science advocacy.

Like the other chapter authors, Marchant believes that there is crisis in regulatory science, in that just as the role of science in government decisionmaking has grown, so too has credibility in regulatory science severely declined. Marchant recognizes that regulatory science is an adversary process, in which regulators marshall the scientific evidence in favor of regulation while interest groups opposed to regulation attack the supporting scientific work with conflicting scientific work. And he recognizes that in many areas of regulation, each side can indeed find at least some supporting scientific work. In his view, on the regulatory side, the problem arises when a single regulatory agency, such as the EPA, is responsible both for assessing the scientific basis for regulation and promulgating and defending regulation.

In the area of endangered species regulation, both Ramey and Wyman agree that by making ESA coverage and legal protection depend upon a taxonomic decision, the ESA has incentivized science advocacy in the field of taxonomy. Ramey argues that once the ESA's protections have been invoked, a well-organized coalition of academics, environmental activists, professional societies, and government regulators and biologists oppose any revision of taxonomic classifications in light of new scientific evidence. The U.S. Fish and Wildlife Service, the agency that implements that ESA, does sometimes utilize peer review in cases of contested taxonomic revisions. But according to Ramey, USFWS peer review is rife with conflicts of interest. The typically small number of experts on a particular species stand only to gain from listing under the ESA, which not only protects their professional specialty but also promises lifetime employment as consultants on how to best protect the species. As Ramey illustrates with the coastal California gnatcatcher case study, USFWS peer review of taxonomic revisions is now often done in cooperation with yet another government agency, the Biological Research Division of the U.S.

Geological Survey. In the coastal California gnatcatcher case, the panel that reviewed the taxonomic determination included no fewer than seven federal government biologists, and, according to Ramey, outside reviewers were not given all the underlying data and information, nor were they allowed to actually perform additional analysis.

Wyman usefully points out that the incentives for science advocacy created by the way taxonomy triggers the ESA's protections do not just apply to the USFWS and to groups in favor of ESA protection, but also to land developers and others who oppose listing. As she argues, just as scientists whose scientific view favors finding subspecies on the basis of even slight genetic variation are natural advocates for those in favor of ESA listing, so too are scientists who believe that subspecies categorization requires much greater difference natural advocates for the opponents of listing.

Johnston and McKitrick focus their attention on the IPCC, which unlike EPA is not a regulatory agency but an international scientific assessment body convened under the auspices of the United Nations. According to both Johnston and McKitrick, the IPCC's structure and process made almost inevitable its evolution into an institution of science advocacy. Under the process by which the IPCC produces its Assessment Reports, politically chosen members of the Bureau select leaders of the Working Groups who in turn choose the Lead Authors and Authors of the Report chapters. While there is indeed an outside review process, Authors are free to ignore the comments of outside reviewers and there is no opportunity for outside reviewers to comment on the final Report. McKitrick effectively debunks the oft-repeated claim that IPCC Assessment Reports are subject to review by thousands of scientists, detailing how for the key chapter on attribution of climate change to anthropogenic activities (chapter 9 of the 2007 Assessment Report by Working Group I, Climate Change Science), there were only thirty-one truly independent outside reviewers. In Johnston's characterization, the IPCC's process gives monopoly authority to Assessment Report chapter authors to certify their own preferred hypotheses as true while rejecting or ignoring conflicting work; in McKitrick's, the failure to acknowledge the self-interested bias of experts has led, in the case of the IPCC, to a system which is analogous to a legal proceeding in which judges appoint advocates for a particular side to make the decision and write the judgment, with only a nominal procedure for the examination of evidence, and with such examination followed by private presentation of additional evidence to judges who have an unlimited right to strike evidence from the record.

THE IMPORTANCE OF BASING REGULATORY SCIENCE ON A BETTER UNDERSTANDING OF SCIENCE AND THE SCIENTIFIC PROCESS

As noted by several authors in this volume, the U.S. Congress has passed a number of statutes which, like the National Ambient Air Quality Standards under the Clean Air Act or the "best available science" mandate for listing species under the ESA, seem to assume that science can provide clear and definite answers and essentially determine regulatory policy. As the chapters in this volume demonstrate in considerable detail, this vision of regulatory science is completely at odds with the reality of how science and the scientific process actually works.

In his contribution to this volume, for example, Ramey begins by making clear what biologists have long known but what many people, including likely the U.S. Congress, do not know: that there are no rules in taxonomy as to what constitutes a species but instead a plethora of as many as twenty-seven alternative definitions, with no consistent thresholds for applying them within or across different kinds of organisms. Given the multiplicity of "species" definitions, it is unsurprising to find, as Ramey clearly describes, that there have been pronounced historical trends in the number of species and subspecies. Through the mid-twentieth century, many new subspecies were declared based on often minor superficial traits or based on small sample sizes, with the peak of such subspecies splitting reached in 1959 with the declaration that there were 214 subspecies of southern pocket gopher. More recently, as the definitions of species and subspecies changed, many subspecies have been elevated to the level of species, a process now known as taxonomic inflation. As Ramey explains, in some of the most notorious cases of taxonomic inflation of legal relevance under the ESA, biologists have used the easy availability of DNA sequencing data to infer that a population is reproductively isolated and therefore a new species or subspecies solely on the basis of tiny genetic variations of no known behavioral, morphological, or reproductive relevance. Similarly, in a number of recent cases, recent geographic separation among populations, due often to twentieth-century land development, has been used to infer reproductive isolation; indeed, in one notorious case involving the declaration of three new subspecies of beach mice, post-World War II dredging of the intracoastal waterway by the U.S. Army Corps of Engineers was found to have created geographic isolation among beach mouse populations.

The "species" concept in biology may well be an extreme case of scientific indeterminacy, but the lack of clear scientific answers to policy-relevant questions is a much more general problem for regulatory science. Johnston and McKitrick's chapters provide many illustrations of areas of active scientific disagreement and controversy on the frontiers of climate science. Conrad similarly provides many examples of chemicals for which there is true scientific uncertainty over the mechanisms and exposure pathways that determine the potential risks to human health. The general problem, succinctly put by Conrad, is that the "best available" science often does not provide the clear answers that regulatory policymakers are hoping to find. Indeed, as a number of chapter authors explain, at the frontiers of any science—whether biology, climate science, or toxicology—one typically finds not the neat consensus that the policymakers would like, but rather ongoing battles among competing methodologies, hypotheses, and evidence. The scientific process—the process by which true and important hypotheses are distinguished from those that are false or unimportant—is imperfect and often halting. As a number of authors point out, peer review publication is only a very coarse screen in which reviewers generally do not have access to the data underlying important results. Whether in climate science, toxicology or taxonomy, the verification of results claimed in published work generally occurs only over time, as other researchers attempt to build upon and extend published results. Contrary to the message conveyed by the popular media, peer review publication is merely the beginning of the process of scientific vetting, not a final certification of true and important work.

The failure to base regulatory science policy on a realistic understanding of the scientific process and the limits of scientific understanding is itself a major contributor to one of the most important but heretofore little recognized problems with regulatory science: the intrusion of policy preferences into science. Ramey and Wyman agree, for example, that by making legal protection depend upon categorization as a "species" or "subspecies" when these concepts in fact have no agreed-upon scientific meaning, the ESA has effectively made the policy preferences of particular scientists the driving force in ESA listing. Wyman's example of this, the controversy over the delisting of the Prebbles jumping mouse in Wyoming and Colorado which occurred during the mid-2000s, is only one of many such cases that have arisen under the ESA. Ramey and Wyman agree that alternative scientific views of whether a particular population of organisms is, for example, a "subspecies" are not just about what the evidence shows: given that there is no scientifically agreed-upon threshold for the amount of genetic difference or geographic isolation required to earn separate categorization as a "subspecies," a biologist's opinion on this question will inevitably reflect her underlying policy judgment about the degree of genetic uniqueness that should be present before a population is legally protected.

One of the most important contributions of the present volume is indeed to reveal the importance of this kind of advocacy science—where a scientist's policy judgments are in fact driving her scientific approach and results. McKitrick emphasizes how the inevitable self-interest of climate scientists in favoring work that supports their own scientific views may bias any purportedly objective scientific assessment. Going further,

Johnston argues for a more direct influence of policy on the supposedly neutral scientific assessment done by the IPCC. On Johnston's view, the IPCC's Assessment Reports have systematically favored work at the frontier of climate science that supports the IPCC's dominant policy position that human greenhouse gas (ghg) emissions have caused and will cause even more harmful global warming unless they are rapidly reduced, while those Reports have systematically discounted or discredited work that casts doubt on these results and supporting hypotheses. Conrad's detailed discussions of ongoing conflicts between industry and the EPA or NIEH over the harmfulness of low-level exposure to substances such as formaldehyde similarly depict an environment in which the policy preferences of scientists on both sides may be expected to drive everything from study design to the publication and interpretation of uncertain results.

Thus an important implication of many of the chapters in this book is that when regulatory policy is based on science that is uncertain, policymakers invite scientists to let their own policy preferences dictate how they do science. If conflicting scientific results are in fact directly driven by conflicting policy-driven choices of everything from methodology to the interpretation of data, then the separation between science and policy, and confidence in the potential neutrality of science, is completely lost.

IMPACT OF SCIENCE ADVOCACY ON THE CONTENT AND CREDIBILITY OF REGULATORY SCIENCE AND THE CONTENT OF REGULATION

A common assumption in discussions of the relationship between science and regulation is that regulatory policy should be based upon an objective assessment of the relevant science. By identifying what may be called the endogeneity of regulatory science, however, several of the contributions to this volume question whether this is possible, at least in the current institutional environment. For purposes of this discussion, regulatory science is endogenous when the body of science that regulators look to is in fact directly shaped by the current legal and regulatory environment. Such endogeneity is of huge consequence, for when there is strong endogeneity, and scientific results are strongly determined by regulatory institutions and policy, then clearly regulations cannot claim to be based on neutral and objective science.

While perhaps all contributors to the present volume would agree that our current regulatory institutions have made at least some regulatory science endogenous, Johnston, McKitrick, and Ramey argue most directly for the endogeneity of regulatory science under current institutions. In the area of climate science and policy, McKitrick provides concrete examples of how IPCC Lead Authors actively influence the peer-edited literature that they are ostensibly objectively assessing for policymakers. McKitrick documents cases where IPCC Lead Authors both cite and discuss work that confirms their views even if it has been rejected for publication in peer-reviewed journals, while at the same time—in their capacity as referees for peer-reviewed journals—attempting to stop the publication of work that disputes their views, and when that fails, ignoring it in their Assessment Report chapter. Ramey describes the close relationship that has evolved between the species regulator, the USFWS, its government scientific advisor, biologists at the USGS, the peer-edited scientific literature where new taxonomic classifications are first claimed, and the professional societies that sign off on taxonomic classifications. What Ramey calls "species cartels" represent the direct influence of ESA-driven regulatory policy on the science of taxonomy.

Drawing on earlier work by Turner (2002), Johnston's contribution sets out an economic model of the scientific process as a decentralized, competitive certification market. He argues that by imposing a centralized, bureaucratic and intentionally political decision about what work on the frontier of climate science is true and important and what work is likely not true and therefore to be discounted, the IPCC has acted to cartelize and politicize climate science. Johnston argues that for policymakers and other lay users of science, what creates confidence in science is the highly decentralized and competitive process by which scientists advance and test hypotheses. For it is only such competition that provides an assurance to users of science that the scientific process has successfully overcome the inherent conflict of interest that is present in any system that relies on self-certification: the incentive for self-interested scientists to collude to certify their own work as true and important and to discredit conflicting scientific work. There is abundant evidence, Johnston argues, that the IPCC has already acted to disturb the otherwise decentralized and competitive scientific process by facilitating collusion by lead IPCC scientists in pressuring editors of peer-edited journals to quickly publish their own work and to suppress or delay the publication of conflicting work, to publish rebuttals to conflicting work much more quickly than would otherwise be normal, and in threatening to boycott journals that do not comply with their demands. Johnston argues that the IPCC has succeeded perhaps too well, for inasmuch as public confidence in climate science depends upon a perception that the climate science process is fair in the sense of being a decentralized, competitive process where true hypotheses win and false ones lose, the IPCC-led cartelization of climate science has harmed the credibility of climate science. Moreover, the IPCC Reports have failed to produce the kind of nuanced and balanced picture of climate science that is necessary to craft sound and sensible climate policy.

PROPOSALS FOR REFORM

All of the contributors believe that the current institutions for regulatory science need to be reformed. The contributors' proposals for how to reform the regulatory science process are quite diverse and range from Johnston's relatively terse recommendation that the IPCC should be abolished and replaced with a system of decentralized, ad-hoc climate science review to Marchant's recommendation for a new U.S. Institute for Scientific Assessment (ISA). For Johnston, who is perhaps the most leery of any centralized institution for scientific assessment, the lesson from the 2010 InterAcademy Council Report on the IPCC's process is that the IPCC process cannot be fixed, and so the IPCC should be abolished. Of all the contributors, Marchant is perhaps the most optimistic in believing that one can design institutions for regulatory science that eliminate the endogeneity problem and restore something like a strong separation between science and policy. According to Marchant, there have been successful models of unbiased and neutral scientific assessment institutions: an expert panel convened by a federal trial court judge in the silicon breast implant litigation in the late 1990s that produced an assessment of the existing scientific evidence on the health impacts of such implants; the European Food Safety Authority, which has now produced over five hundred assessments of food safety in response to requests from various European regulatory bodies; the Health Effects Institute, a joint EPA-auto industry funded group that provides assessments of the state of the scientific evidence on the health effects of regulated air pollutants; and, finally, the IPCC, which Marchant believes can accurately represent the disparate scientific views in climate science with some structural changes and safeguards to minimize the mixing of policy and science that has recently harmed the IPCC's credibility. Marchant's proposed ISA would focus just on the evaluation of scientific data and issues. It would provide a "credible independent assessment" of the state of the relevant science and thus make it much harder for regulatory agencies to camouflage their policy preferences as science.

Ramey's reform proposal also reflects some degree of confidence in the ability to separate science from policy. He believes that what has caused most of the taxonomic problems in ESA listing is the failure to have a uniform standard setting a threshold level of uniqueness that would need to be proven for a population to qualify as a subspecies or species under the statute. Ramey sets out a number of interesting possibilities for setting such a threshold, including threshold quantitative measures of statistically significant genetic uniqueness, prioritization based on existing quantitative measures of evolutionary distinctiveness and level of endangerment, and even more inclusive quantitative methods of triage across species.

Wyman is more skeptical of the ability to reform the ESA so as to effectively disentangle policy from science. In her view, there is inevitably scientific uncertainty in any determination of a threshold level of uniqueness for legal action, and the choice of how to treat such uncertainty inevitably involves the policy decision about how to weigh the value of saving species against the cost. Wyman advocates approaches which would reduce the legal significance of scientific determinations of both taxonomic category and extinction risk. She builds upon her own previous proposal that listing should be followed by a fine-tuned, cost-sensitive approach to designing permanent species protections (versus the current one-size-fits-all approach now found under the ESA), by recommending that additional permanent protected areas for species on public lands might be used to provide un- or de-listed species with additional safeguards, thus lessening the need to list and the cost of listing to private landowners.

Dudley and Gray stress reforms that tend to focus on what has become known as the "science charade," the tendency for regulators to mask policy decisions as dictated by science. For this reason, Dudley and Gray urge reforms intended to force regulators to clearly distinguish between regulatory decisions that involve science and those that involve policy. Their recommendations include: having agency scientists first calculate and present the multiple risk assessments, and then separately identify the agency's preferred estimate; making public the science and science policy choices in regulatory science assessments; requiring that agencies make public external peer reviews of agency scientific assessments and make publicly available also the agency responses to those comments, with a further opportunity for peer reviewers to respond to the agency's response; and, finally, they suggest that comments on rulemaking might be done in an electronic "wiki" type format, where there is ongoing collaboration in refining comments, rather than separate and litigation-oriented comments submitted in parallel.

Conrad's proposed solutions combine elements of both more definite scientific standards and more open and transparent policymaking processes. On the one hand, he argues that there do exist generally agreed-upon methodologies in toxicology and epidemiology for evaluating both individual as well as multiple studies. By adhering to such methodologies, agencies such as EPA would minimize unnecessary disputes involving the interpretation of existing work, and would also commit themselves not to engage in the practice of cherry-picking a large literature to find those few studies that support a finding of significant risk at low doses. In those situations where there is true scientific uncertainty over the mechanisms, exposure pathways, or other science relating to the riskiness of a chemical, Conrad recommends that agencies like EPA hold early symposium-like proceedings in which interested parties, peer reviewers and agency staff—who would all have had adequate time to review draft risk assessments—would meet in a setting where peer reviewers and agency staff could be questioned by interested parties, who would have an opportunity to fully inform regulators about new and relevant scientific work that might conflict with the agency's preferred findings. Finally, he recommends a more continuous and adaptive risk assessment process, with agencies viewing all estimated risks as essentially provisional and subject to change in light of new scientific findings.

In McKitrick's view, institutions for scientific assessment must be based on the recognition that experts will "typically have preferred points of view or convictions on matters related to their areas of expertise." Given the preferences and self-interest of experts in endorsing work that supports their own, one cannot expect, McKitrick argues, that members of expert panels will somehow be able to survey their own fields in an unbiased and neutral way. Unlike, for example, Marchant, McKitrick does not find that existing expert panels and other existing institutions for scientific assessment effectively overcome this inherent conflict of interest. McKitrick finds fault with National Academy of Sciences (NAS) expert panels, for while NAS guidelines for panel selection require that panels should be "balanced," they also caution against advocacy. Yet in McKitrick's view, advocacy is inevitable when scientists are asked to choose between their own and competing views. Indeed, in his own experience in presenting to an NAS panel on the so-called "hockey stick" controversy involving reconstructions of past global temperatures, McKitrick found that the panel was made up primarily of people who had publicly supported the IPCC's use of the "hockey stick" data to dramatize climate change and who therefore had no interest in allowing that work to be seriously disputed or cross-examined. In the end, McKitrick argues that institutions like the IPCC should be reformed to allow what amounts to an adversary form of presentation, with conflicting sides in ongoing scientific disputes allowed to make their case as best they can.

REFERENCES

McGarity, Thomas O., and Wendy E. Wagner (2008). *Bending Science: How Special Interests Corrupt Public Health Research*. Cambridge, MA: Harvard University Press.

Turner, Stephen (2002). "Scientists as Agents." Pp. 362–83 in *Science Bought and Sold*. Mirowski, Philip, and Esther-Mirjam Sent, eds. Chicago: The University of Chicago Press.

Wagner, Wendy E. (1995). "The Science Charade in Toxic Risk Regulation." *Columbia Law Review* 95(7): 1613–1723.

Wagner, Wendy E., and Rena Steinzor (2006). *Rescuing Science from Politics: Regulation and the Distortion of Scientific Research*. Cambridge: Cambridge University Press.

I

Institutions for Climate Science Assessment

TWO

The Cost of Cartelization

The IPCC Process and the Crisis of Credibility in Climate Science

Jason Scott Johnston[1]

INTRODUCTION

The Intergovernment Panel on Climate Change (IPCC) was established in 1988 as a joint subsidiary of the United Nations Environmental Program (UNEP) and the World Meteorological Organization (WMO). The IPCC's original stated mission was to assess ". . . on a comprehensive, objective, open and transparent basis the scientific, technical and socioeconomic information relevant to understanding the scientific basis of human-induced climate change. . . . " (IPCC 2011). The main outputs of the IPCC process are its periodic Assessment Reports, produced by three topical Working Groups—climate science, climate change impacts and adaptation, and climate change mitigation. As described by a defender of the IPCC process, ". . . in terms of impact on the global political agenda, [IPCC] consensus reports remain entirely without precedent in the policy-making process" (Zillman 2007). Indeed, twenty years after its creation, IPCC influence was evidenced by a (collective) Nobel Peace Prize.

The trust that so many people have in the IPCC's substantive conclusions about climate science seems based very largely on their confidence in the process by which IPCC Reports are produced. For example, the admirably rigorous American legal scholar Daniel Farber (2007, 5) has argued that confidence in the IPCC Reports is justified because those reports are the outcome of an "exhaustive review process" involving hundreds of outside reviewers and thousands of comments. As the British economist David Henderson (2007a, 195) has surmised:

> it is in the network, and the reporting process which brings its members together, that trust is placed: people visualize an array of technically competent persons whose knowledge and wisdom are effectively brought to bear through an independent, objective and thoroughly professional scientific inquiry. Indeed, many people make no distinction between the network and the Panel itself, as though well-qualified and disinterested people were the only people involved.

Leaders of the IPCC have worked diligently to support this image of the IPCC process as neutral, unbiased and including literally thousands of climate scientists from around the world. The IPCC's longstanding Chairman, the energy economist R. K. Pachauri, for example, has said that the IPCC:

> mobilizes the best experts from all over the world, who work diligently in bringing out the various reports of this body on a regular basis. The Third Assessment Report (TAR) of the IPCC was released in 2001 through the collective efforts of around 2000 experts from a diverse range of countries and disciplines. All of the IPCC's reports go through a careful two stage review process by governments and experts and acceptance by the member governments composing the panel.[2]

Recently, through examination of the stolen Climategate emails, this perception of the IPCC process has begun to change. Those emails reveal active and ongoing disputes at the frontier of climate science, disputes in which climate scientists who have been leaders in the IPCC have fought, sometimes viciously and often uncivilly, to defend their claims about what is known and true in climate science against rival and conflicting claims made by other climate scientists. Some of the most colorful of these fights have taken place in the context of peer-reviewed climate science journals, where the Climategate emails show leaders of what I shall call establishment climate science attempting to suppress and discredit work that disagrees with their own.

Spurred in part by the Climategate emails, a number of people have begun to take a closer look at the IPCC Assessment Reports, the most recent of which were published in 2007, and have found that some of the IPCC's most disturbing and most prominently publicized claims—such as the projected loss of Himalayan glaciers by 2035, twenty-five years from now—are in fact not accurate statements of what one finds in the peer-reviewed scientific literature that the IPCC was charged to review and assess. More generally, as I have argued in related work (Johnston 2010), when one compares the peer-edited climate science that the IPCC's 2007 Assessment Reports (AR's) endorsed with the actual literature, one finds a systematic tendency for the IPCC's Reports to engage in a variety of rhetorical techniques that seem much more characteristic of advocacy than unbiased and objective assessment: neglecting to mention work that conflicts with or otherwise weakens work that the IPCC endorses, a systematic tendency to minimize the existence and importance of remaining open scientific questions, failure to mention key assumptions that drive climate model projections and which allow climate models to reproduce past temperature trends, and reliance—especially in the Summaries for Policymakers which are admittedly the basis for media stories and popular perception—on numbers for things such as species loss that have been generated by methods many scientists deem to be invalid. All in all, there is more and more evidence that the IPCC's search for "consensus" in climate science has instead turned into a process where the IPCC has become not an unbiased assessor, but an active advocate for some claims on the frontier of climate science and an active opponent of others.

Such active advocacy is not, I believe, what the IPCC has been advertised as doing. In this chapter, I begin by describing various pieces of evidence that all indicate that the IPCC has become an active advocate of scientific work that tends to confirm the view that there is now strong evidence that anthropogenic warming has already begun and will become much worse, while it actively opposes any scientific work that tends to disconfirm this established view in whole or part.

After setting out evidence that there is indeed a problem with the IPCC acting as advocate rather than unbiased assessment body, I then explain how the IPCC's structure and process has made this problem inevitable. One defect with the IPCC structure is indeed, as Roger Pielke Jr. (2006) has put it, that the IPCC's " . . . institutional organization, selection of participants, and even scientific foci necessarily reflect a non-neutral policy orientation, and hence it is very political and advocacy oriented." As a direct reflection of its advocacy orientation, the IPCC assessment process gives virtually unbridled discretion to Working Group chairs and authors of IPCC Assessment Reports to identify "correct" versus "incorrect" work on the frontier of climate science. As concluded by the 2010 InterAcademy Report on the IPCC process, there is no existing institution that could provide the objectivity necessary to overcome the conflicts of interest that are inherent in the IPCC process as currently structured. The IPCC process will inevitably privilege one side in areas of active and ongoing dispute at the frontiers of climate science. Given the IPCC's structure and process, climate science that confirms the policy view that human greenhouse gas (ghg) emissions should be rapidly reduced will always win out in IPCC Assessment Reports.

The IPCC has had a harmful impact on both climate science and climate policy. As for climate science, the IPCC process had disrupted the highly decentralized and diffuse process by which claims to new scientific knowledge have traditionally been filtered and certified for truth and importance—a process in which the traditional "scientific method" plays the primary role—with a centralized, bureaucratic endorsement process dominated primarily by the perceived policy consequences of scientific truth or falsity. Leading IPCC climate scientists have collectively intervened in the peer-reviewed publication process in attempts to suppress entirely or minimize the impact of scientific work that questions or conflicts with the IPCC-endorsed view of climate science. These efforts represent an attempt to cartelize the otherwise competitive market for scientific certification. Such cartelization has lessened the credibility of climate science.

As for climate policy, the IPCC Reports have not provided the balanced and detailed assessment of climate science that is required for a sophisticated and sensible climate policy. Instead, through a single-minded focus on human CO_2 as the primary cause of recent and projected climate change, and through misleadingly confident projections of potential future catastrophic harm if human ghg emissions are not quickly reduced, the IPCC Reports have provided only very coarse information for policymakers. Even worse, those Reports have tended to provide justification for even the most costly and crude climate change policy responses (such as banning the use of coal).

The IPCC cannot, I conclude, be fixed. It should be abolished. While policymakers do indeed need up-to-date assessments of climate science, in designing an institution to provide such advice, much work is required in identifying how alternative institutional structures impact the incentive of scientists to provide fair, balanced, and understandable descriptions of both what is known and unknown in climate science. In the meantime, policymakers are better served by funding ad hoc, decentralized assessments of the state of climate science than by continuing to fund the IPCC.

I. THE IPCC AND ITS ASSESSMENT REPORTS: EVIDENCE OF ADVOCACY, NOT UNBIASED ASSESSMENT

In this section, I provide evidence that the IPCC Assessment Report process is one in which the IPCC attempts to persuade people across the world that certain work on climate change is true and important while other work—itself published in peer-reviewed journals of equal rigor and quality—is not true and should not be credited. Such advocacy would not be troubling were the IPCC a party in adversary civil litigation. However, the IPCC is not supposed to be a party in litigation, but an unbiased and objective international assessment institution sponsored by a United Nations agency. If, as I show below, it is in fact acting as an advocate, then there is serious risk that climate change policy will be based on a distorted view of climate science.

There are many pieces of evidence that the IPCC acts as an advocate rather than as an unbiased assessor. This has been the express conclusion, for example, of some of America's most prominent climate scientists, including Richard Lindzen of MIT and Christopher Landsea of NOAA. These scientists once actively participated in the IPCC process (as, respectively, a lead author, and author and reviewer), but have terminated their involvement with the IPCC and did not participate in the process that produced the IPCC's 2007 Assessment Report (see Landsea 2005). Below I discuss some of the abundant additional evidence of the IPCC as advocate that confirms these first-hand reports by former participants. First, a brief summary comparison of the IPCC's version of climate science with the peer-edited climate science literature reveals that on virtually every major claim in the established climate story, there is contested work that the IPCC Reports minimize, dismiss, or simply fail to even acknowledge. Second, as for the IPCC publication process, there is a deliberate IPCC strategy of releasing the popularly broadcast Policymaker Summaries well in advance of the full Reports to which they cite; such early release of Policymaker Summaries strongly contributes to a biased popular impression of the state of climate science because the Summaries systematically fail to portray the full range of uncertainty and qualifications that one finds in the peer-edited literature (and sometimes in the full Report chapters themselves). A final piece of evidence of advocacy comes from the Climategate emails. Those emails depict lead IPCC scientists engaging in systematic attempts to suppress the publication in peer-edited journals of scientific work that disputes views of climate science that are endorsed by the IPCC.

A. Evidence of Advocacy, I: The IPCC versus the Peer-Reviewed Literature

Here I summarize some of the main findings from a detailed comparison of what one finds in IPCC Reports with what one finds in the peer-edited climate science literature (Johnston 2010). In such a comparison—essentially the kind of cross-examination that an attorney conducts when deposing an opposition expert witness—one asks whether the IPCC has fairly summarized the state of the literature, or has instead failed to acknowledge disputes over methodological validity and data reliability, conflicting studies, and studies that tend to indicate that the establishment story is not so clear cut as the IPCC Reports now portray it to be. In related work, I have undertaken such a cross-examination (Johnston 2010). Here, I summarize its main findings.

The establishment climate story—the vision of climate science endorsed by the IPCC—consists of at least the following basic claims: 1) as the concentration of ghg's in the atmosphere increases, global surface temperatures (as well as temperatures in the troposphere) will increase; 2) powerful positive feedback effects—such as increased atmospheric water vapor and melting snow and ice—will amplify the temperature increases; 3) while there is no way of knowing for sure, temperature increases could be very high, as much as 4 degrees Celsius or more from a doubling of atmospheric CO_2; 4) we are increasingly certain of all of these things because temperatures have already been increasing, reaching levels that are very unusual in earth history, and this increase has been due to increasing ghg concentrations in the atmosphere; 5) we know that the recent increase in temperatures is due to increasing atmospheric ghg concentration because climate models allow us to pinpoint these temperature increases as a result of increasing concentrations of ghg's, not something else, such as solar variation or natural climate variability; 6) climate models are to be trusted because they quite accurately are able to reproduce past temperature trends, especially twentieth-century temperature trends; 7) even if we didn't have climate models, paleo-climatic evidence is very reliable and shows that late twentieth-century temperature increases are really unusual in earth history; 8) such paleo-climatic evidence also shows that if ghg concentration continues to increase and temperatures increase to the levels that climate models predict as a consequence, then there will be massive and potentially catastrophic changes, such as increasing sea level, and even possibly shutdowns of major ocean current systems; 9) reliable models from disciplines such as ecology and epidemiology show that if temperatures increase as climate models predict, then there will likely be massive die-offs of non-human species and massive increases in human suffering due to increased prevalence of climate-related diseases; and, finally, 10) the only real response to these potential harms is to make large investments today to reduce human ghg emissions.

Especially in the Summary for Policymakers released with its recent 2007 Report assessing climate science, the IPCC has stressed that the accumulating body of scientific work supports all ten of these claims, and with increasing confidence. My cross-examination revealed to the contrary that on virtually every one of these major claims, the peer-reviewed scientific literature paints a much more nuanced, evolving and uncertain picture. Consider what one finds in the peer-edited literature with respect to the following major claims made by the IPCC:

1) There is substantial evidence (see, e.g., Pielke Sr. et al. 2006; McIntyre & McKitrick 2005) that both twentieth-century and longer run temperature estimates endorsed by the IPCC are biased and both overestimate late twentieth-century temperature increases and underestimate temperatures during earlier epochs in earth history;

2) The IPCC has projected that it is "very likely" that a doubling of atmospheric carbon dioxide will induce a 1.5–4 degree centigrade increase in global mean surface temperature (IPCC 2007, 749) (temperature projections from climate models are by convention for a doubling of CO_2 and so this will be assumed in future references). The peer-reviewed climate science literature shows that projected increases above 1 degree C are due entirely to positive feedback effects presumed by climate models

(Roe & Baker 2007), yet while feedback effects are discussed in some detail in the 2007 Report's chapter entitled "Climate Models and their Evaluation" (IPCC 2007, chapter 8), feedbacks are given almost no mention in the Summary for Policymakers that is (as I explain below) so influential with the media, the public, and policymakers;

3) Buried deep in the IPCC's 2007 Report (chapter 8, which begins at p. 589) is a discussion justifying climate models' derivation/assumption of strongly positive feedback effects, but there is published evidence that the climate models relied upon by the IPCC Report are underestimating negative feedback effects (such as low cloud albedo) and overestimating positive feedbacks (most importantly, the water vapor feedback) (see Sun et al. 2009); more generally, the existing literature reveals that there is very little empirical support for the IPCC Report's crucial conclusion that feedback effects are on balance strongly positive (see, e.g., Wang et al. 2008);

4) Theoretical work shows that the presumption of net positive feedbacks itself mathematically generates a positive probability of very large temperature increases, but that if there is a significant negative feedback (but one that dissipates over time) then it may be centuries before such big temperature increases occur with positive probability (Baker and Roe 2009);

5) Although barely mentioned by the IPCC, recent work shows that a substantial fraction of the twentieth-century Arctic warming that is claimed by the IPCC to evidence CO_2-driven warming is in fact due to black carbon or soot emissions (Flanner et al. 2007), and scientists continue to publish papers reporting evidence that solar variation and/or natural climatic cycles may account for a substantial fraction of overall late twentieth-century warming (see the summary of such papers in Scafetta & West 2008);

6) Moreover, it has recently been shown that climate models—which disagree enormously on the sensitivity of temperature to increases in CO_2 concentration—are nonetheless all able to "reproduce" twentieth-century temperature trends only by making widely varying and seemingly arbitrary assumptions about the net cooling impact of atmospheric aerosols (Schwartz, Charlson & Rodhe 2007); this crucial limitation of climate models is nowhere clearly acknowledged or explained in the IPCC Report;

7) A number of studies cast doubt on the ability of paleoclimatic studies to identify the direction of causality between past increases in atmospheric CO_2 and global surface temperature (see, e.g., Stott et al. 2007), and when it comes to sea level rise, there is wide variation in estimated increases over the twentieth century (see the summary of various estimates presented by Khandekar 2009), and some scientists doubt that sea level rise can be accurately estimated given current methods (Wunsch et al. 2007);

8) A large number of biologists have criticized the models that the IPCC relies upon for its projections of species loss from global warming as methodologically invalid and as almost sure to overestimate future species loss (Lewis 2006; Botkin et al. 2007), yet the Summary for Policymakers in the IPCC's 2007 AR on climate change impacts (IPCC 2007a, 8) broadcasts the species loss prediction without any qualification whatsoever regarding the highly controversial and questionable validity of the models that generate those numerical predictions.

The cross-examination thus reveals that what has emerged from the IPCC process is a Report, and popularly-targeted Report Summary, that culls the existing peer-edited literature for work that supports the established climate story, and which presents such work as the "truth" about climate change. When the IPCC Report does discuss work that conflicts with work it endorses, the discussion does not often reveal the actual substance of the dispute. Importantly, the IPCC tends to gloss over serious existing disputes over the reliability of competing observational datasets. More often than not, the IPCC Report simply flat out ignores work that weakens the established story, even though that work has been published in top peer-edited journals. This kind of rhetorical technique—in which some literature is endorsed while other literature is mentioned in only the most cursory way or simply ignored—makes it extremely costly to conduct even a cursory audit of the IPCC Report. Even more seriously, this rhetorical strategy seems very likely to create a very serious misimpression about the state of frontier climate science, portraying much more agreement among climate scientists on key questions than is actually found in the literature. It seems clearly to be a rhetorical strategy designed to persuade the reader that certain climate scientists are "right" and others "wrong" in areas of ongoing scientific investigation at the frontier of climate science, rather than to paint an accurate picture of ongoing work.

B. Evidence of Advocacy, II: The IPCC's "Summary for Policymakers" Release Strategy

Before the publication of each Assessment Report by the various Working Groups, the IPCC issues what it calls a "Summary for Policymakers"—relatively short documents that purportedly summarize full Assessment Reports—several months prior to the issuance of full Assessment Reports. Yet these "Summaries" in fact often completely fail to mention key qualifying and explanatory conclusions in the full reports. Even more seriously, the Summaries contain numerous supporting citations, but these citations are to the full Assessment Reports, which are typically not made accessible online until months after the Summaries appear. As I illustrate with some examples below, the Summaries contain a rhetoric of certainty that often turns out to be flatly contradicted by the language of the main Report. This contradiction, and the long delay between the publication of the Summary for Policymakers and the Report that it relies upon as authority, seem deliberately intended to disable and weaken critical analysis.

Criticism is disabled because by saying one very clear thing in the Executive Summary and another, much longer and more highly qualified thing in the Report, the IPCC confronts the critic with a dilemma: one cannot say that the IPCC really did not say "x" when it said "x" clearly in the Summary and then not so clearly said it did not really mean "x" in the full Report. This IPCC rhetorical strategy sets up what seems to be a deliberate contradiction between what has been released to the media and policy makers and what is being told to more sophisticated audiences.

This "acoustic separation" is especially effective because of the long delay—typically of several months—between the release of a Summary and the release of the full Report being summarized. Although Summaries cite to the full Reports, it is impossible to verify any claim or assertion in a Summary until months after one is made public. By the time a full IPCC Report is released, the popular media has long since broadcast the "findings" presented in the Summary as identical to and supported by the full IPCC Assessment Report, so that it is the Summary (and not the full Report that it may well misrepresent) that has the first and often lasting impact on popular opinion and public policy.[3]

An example of the influence of the early publications of so-called "Summary for Policymakers" is provided by popular media impact of the "Summary for Policymakers" of the IPCC Working Group I "Physical Science" basis Report. As put by the "In Focus" editorial column of the popular magazine Scientific American, the "debate over whether Earth's climate is changing and if humanity is responsible for that change closed in Paris on February 2. The Intergovernmental Panel on Climate Change (IPCC) released its summary for policymakers—a summation of the salient science in its much longer report due in May—in which it said that climate change is 'unequivocal' and estimated the chances of humans being behind it at 90 percent, or 'very likely.'" (Biello 2007).

An even more striking example of the IPCC's use of Summaries as an advocacy tool is its so-called Synthesis Report, released on November 17, 2007. This short, seventy-three-page Report is the only Report that was issued simultaneously with its accompanying twenty-two-page Summary. While the Synthesis Report might seem an exception to the general IPCC acoustic separation strategy that I have described, it was not. The content of the Synthesis went beyond any previous report, including a new section, entitled "Reasons for Concern," that discussed highly uncertain consequences of global warming, such as increases in the frequency or severity of extreme weather events, and even events—most particularly, rapid melting of polar ice—that while highly alarming had been described as not likely in the previous full technical scientific report. In effect, the Synthesis Report took what the earlier Report had said about polar ice melt rates—that they were not predicted by existing models but that over centuries, continuing warming could cause the loss of most or even all of the Greenland ice sheet—and translated it into the statement that since the models didn't predict current melting, it is possible that melting could occur much faster and sooner than expected. As for its Summary, the Synthesis Report summary was not really a summary; the only difference between the full Synthesis Report and its Summary is the more strident rhetoric of the Summary.

In terms of timing, the Synthesis Report replicated the prior pattern of early release to generate press and public attention. Paralleling the IPCC strategy in releasing Summaries months before full reports, journalists reported that drafts of the Synthesis Report had been circulating "for weeks" before its release (Rosenthal 2007), which itself was timed to occur just before December, 2007 negotiations on climate change in Bali, Indonesia. As publicly characterized by a leading IPCC scientist, "[y]ou look to a synthesis report to provide clarity, to clarify what was obscure in previous reports . . . Now, how can we take these findings and formulate a policy response that's quick enough and big enough?" (Rosenthal 2007). According to climate scientist-advocate Michael Oppenheimer, the Synthesis Report's alarmist rhetoric on ice sheet melting set out a scenario "that would make it not just difficult, but impossible to adapt successfully, some of my colleagues would say catastrophic . . . If they say that it's possible that melting could occur in centuries leading to meters of change, that's a headline" (Rosenthal 2007). Indeed, by stressing the most alarming, albeit highly uncertain, potential impacts of global warming, such as species loss, the Synthesis Report seemed clearly designed, as put by Oppenheimer, to "light a fire under policy makers." As for why earlier reports had failed to emphasize these alarming but highly uncertain consequences, Oppenheimer opined that "governments had 'messed around' with previous reports" (Rosenthal 2007).

C. Evidence of Advocacy, III: The Significance of "Climategate"

In the fall of 2009, thousands of emails were hacked from the server of the University of East Anglia's Climate Research Unit (CRU), a research center that is one of the relatively small number of institutions that generate historical data on global surface temperatures. These emails consisted of years of correspondence among climate scientists who have been leaders in the IPCC process since its very inception. This correspondence reveals leading climate scientists engaging in a variety of behind-the-scenes activities designed to squelch their scientific and political opponents and advance their own views of climate science and policy. The 2009 email release has become popularly known as "Climategate."

One of the most disturbing incidents revealed by the Climategate emails involved the publication of two papers testing one of the few empirically testable predictions generated by climate models—that if global warming is being caused by the buildup of anthropogenic ghg's in the atmosphere, then more warming should be observed in the tropical troposphere than at the tropical surface level. Both papers were published in the International Journal of Climatology. The first paper to be accepted by that journal was Douglass et al. (2007) and the second paper to be accepted was Santer et al. (2008). Douglass et al. (2007) argue that the observed temperature trends in the tropical troposphere differ in a statistically significant way from those predicted by climate models; looking at different data and employing a different statistical test, Santer et al. (2008) argue that there is no statistically significant difference between observed and predicted tropical tropospheric temperature trends.

In an online essay entitled "A Climatology Conspiracy?" Douglass and Christy (2009) have argued that the Climategate emails show that the Santer et al. (2008) paper was in fact a rebuttal to their own 2007 paper, and should therefore have appeared in print as a comment or response to their paper only after their paper was published in print (versus online); the emails show, say Douglass and Christy (2009), that the Santer et al. (2008) paper appeared in the same print version of the journal—with no opportunity for rebuttal by Douglass et al.—only because Santer's coauthors directly lobbied the journal editor to delay print publication of the Douglass et al. paper so that it would be published in print simultaneously with the Santer et al. paper. Santer (2010) responded to the charges of Douglass and Christy (2009) with an essay in which he again attacked the statistical methodology used by Douglass et al. (2007) and rebutted various other allegations made by Douglass and Christy (2009). However, regarding the key ethical charge made by Douglass and Christy (2009)—that Santer and his coauthors conspired together to influence the editor of the International Journal of Climatology to delay print publication of the Douglass et al. (2007) paper—Santer (2010, 10–11) says only that "[t]he decision to publish the print versions of the Douglass et al. and Santer et al. IJoC on the same date was an editorial decision. It was not my decision . . . ," adding that "in the scientific world, most journals now publish papers online before they appear in hardcopy form" and that "the online publication is generally considered to be the publication of record." As Douglass et al. (2007) appeared online 10 months prior to Santer et al. (2008), Santer (2010, 11) concludes that "any (imagined or imaginary) conspiracy to delay publication of Douglass et al. would therefore have to be judged remarkably unsuccessful."

What matters for present purposes is not the substantive disagreement between Douglass et al. (2007) and Santer et al. (2008),[4] but rather the intense advocacy efforts by Santer et al. The Climategate emails are all available online, and they clearly confirm the picture painted by

Douglass and Christy (2009) of a feverish group effort by Santer and coauthors first to stop the publication of the study by Douglass et al. (2007), and then, failing in that endeavor, to influence the editor of the International Journal of Climatology to publish their paper rebutting Douglass et al. (2007) as quickly as possible. I know of no systematic evidence on whether such intense lobbying by such a large group of coauthors is normal in climate science; based on my own experience, it would be extremely unusual in economics or law economics, and would be viewed as an infringement upon the professional norm of editorial independence and expertise. Most troubling of all, and most relevant for present purposes, the Santer coauthor group was not only exceptionally large—consisting of seventeen climate scientists—but was virtually a "who's who" of the IPCC science establishment: Every one of the Santer et al. group has served as either an IPCC AR lead author or Working Group chair. Indeed, so large was the Santer et al. coauthor group that its membership seems to have been dictated more by the essentially political objective of including as many leading IPCC-endorsed climate modelers as possible than by the need to pool collective expertise. Leaving aside the possible question of whether including as coauthors individuals who have not actually contributed to the paper violates climate science norms,[5] the Santer et al. coauthor group effort clearly represents an intentional and coordinated effort by IPCC scientists to directly use their positions as IPCC scientists to influence the process by which peer-reviewed journals make publication decisions.

The significance of the Climategate emails is that they depict a number of episodes which, like the Douglass et al. vs. Santer et al. battle, involve leading IPCC scientists acting as a group to attempt to suppress the publication of contradictory work or to pressure journal editors to quickly publish their rebuttals of such work. Such episodes are significant because the credibility of the IPCC process depends upon its being held out as one in which leading scientists objectively assess the peer-reviewed literature. It seems flatly inconsistent with any plausible model of human behavior or definition of "objectively" to believe that the very same lead IPCC scientists who have been spending years' worth of effort to suppress and/or discredit views that they disagree with can then write a report in which they "objectively" assess such views.

The Climategate emails were far from the first evidence of the IPCC's active engagement in endorsing and defending controversial work in climate science. Perhaps the most well-known evidence of such advocacy is in the IPCC's use and intensive marketing in its 2001 Assessment Report of the so-called "hockey stick" graph. This graph depicted very stable and falling global temperatures over nine hundred of the last thousand years, until temperatures began rising rapidly just before the twentieth century, rising in tandem with rising anthropogenic CO_2 emissions. The graph was based entirely on work done by the then young climate scientist Michael Mann, who was also the Lead Author of the IPCC chapter that used his work. The hockey stick temperature graph was strongly endorsed by the IPCC's 2001 Report, this despite published work revealing severe problems with the data and statistical methods used to construct it, published work that had led to a Congressional inquiry and a very critical National Research Council (NRC) Report. McKitrick's chapter in this volume and Holland (2007) provide additional and dramatic detail regarding the hockey stick affair and the NRC Report process. There seems no doubt that the IPCC has become an institution that gives authoritative weight to some views of contested issues in climate science, weight that IPCC scientists then exert to directly influence the development of the very same body of climate science work that they are supposed to be objectively reviewing.

II. A STRUCTURE AND PROCESS EXPLANATION FOR THE IPCC AS ADVOCACY INSTITUTION

In this part, I argue that the IPCC's structure and process contain certain features that inevitably tend to put scientists who are engaged in active and ongoing disputes at the frontier of climate science in the position of adjudicating, for purposes of IPCC Assessment Reports, who is right and who is wrong in such disputes. On this view, the IPCC's basic process creates an inevitable conflict of interest for Lead Authors and Authors of IPCC Reports. The scientists who write the IPCC Assessment Reports are supposed to report on the current state of cutting edge climate science, but since the cutting edge is contested, these scientists inevitably end up reporting that one side—their side—is correct, while ignoring or dismissing work that disputes their claims.

This is not the view of the IPCC process taken by IPCC leaders, and hence after giving my critique, I present one prominent defense of the IPCC process and the IPCC's role. I conclude by summarizing some of the key conclusions of the recent 2010 review of the IPCC process undertaken at the U.N. and IPCC's request by the InterAcademy Council. The InterAcademy Report's conclusions were not only entirely consistent with my own critique, they also suggest that there is no way to easily fix the IPCC process to eliminate the worst conflicts of interest.

A. The IPCC Process: An Overview

I begin with a brief review of the overall IPCC structure.[6] The Panel (it is called, after all, the Intergovernmental Panel on Climate Change) itself is made up of officials—both scientists and laypeople—appointed by the governments who are signatories to the Framework Convention on Climate Change. It is these government agencies who are effectively the

IPCC Panel's clients. The IPCC—that is, the Panel—meets in plenary sessions. It specifies work plans for the Working Groups of scientists and others who prepare the IPCC Assessment Reports, and provides guidelines for the composition and type of report. Although the Panel has ultimate control over the process by which Assessment Reports are created, it is the Bureau, a group of experts chosen and broadly controlled by the Panel, who more directly influence the Report production process. (As pointed out by Boehmer-Christiansen 147 Bureau members were originally recommended by UNEP and WMO.) Members of the Bureau advise the Panel during small meetings and plenary sessions where draft Assessment Reports, and Summaries of those Reports, are considered, but perhaps the most important role of Bureau members is in working with the IPCC Secretariat and governments and "participating organizations" to select the individuals who actually write and help review the Assessment Reports: the Coordinating Lead Authors, Lead Authors and Contributing Authors who write the Reports, and the outside expert reviewers and internal Review Editors who review the Report (IPCC 2008, sections 4.2.1 and 4.2.2).

The output of this process are IPCC Assessment Reports and summaries that accompany each Report—a required Summary for Policymakers and an optional Technical Summary. Full IPCC Assessment Reports are "accepted" at "sessions of Working Groups," which means that a plenary meeting of the governmental representatives to the Working Group that

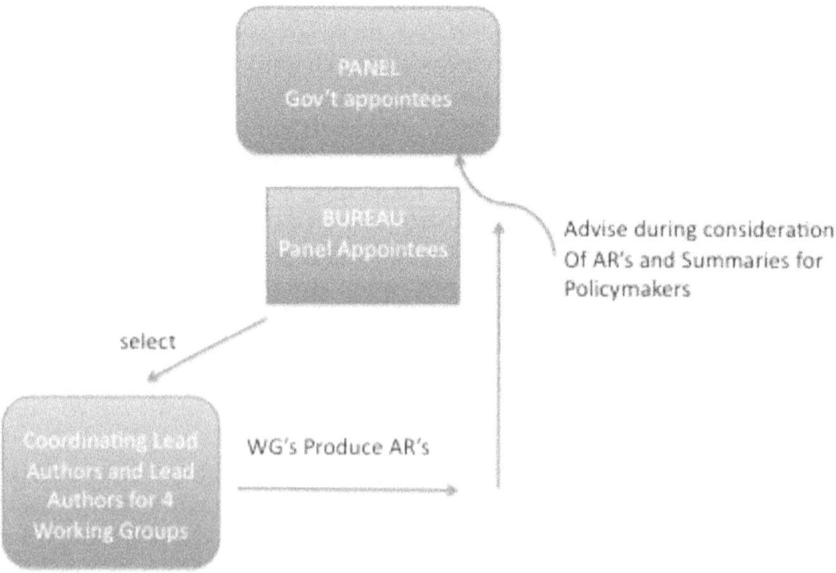

Figure 2.1. IPCC Structure

produced the Report has endorsed the AR as presenting a "comprehensive, objective and balanced view of the subject matter" (IPCC 2008, section 2). The IPCC Working Groups also "approve" Summaries for Policymakers, where "approval" means that such Summaries have been subject to "detailed, line by line discussion and agreement" at such Working Group sessions; such Summaries must also be "accepted" by the full IPCC Plenary (IPCC 2008, Section 2). The basic IPCC process, and its outputs, are described in Figure 2.1 below.

B. The IPCC's Intermixing of Politics and Science

The first thing to note about the IPCC process is that it provides many opportunities for governments to very strongly influence the context of the Assessment Reports and the Summaries that accompany them. The government agencies who are the ultimate clients of the IPCC exert their influence at three key points. First, since they provide the lists from which chapter lead authors are chosen, government actors exert control over the selection of chapter lead authors. As I explain shortly, chapter authors have enormous discretion over the content of their chapters. Second, government client agencies get complete draft texts of the Assessment Reports, and these government agencies must approve the final versions of Assessment Reports. It is at yet a third stage, the production of the very influential "Summaries for Policymakers," however, that government agencies may have their most direct influence. Policymaker summaries are reviewed and then approved, line by line, by governments in a process that one observer has said "often resembles a fox-trot performed by a drunken couple . . . the final negotiated statements from such sessions are often based on least common denominator conclusions written in carefully hedged language . . . the assessments make frequent trade-offs between maintaining scientific credibility and enhancing the policy specificity of their conclusions. . . . " (Agrawala 1998, 627). Through this line by line approval process, government agencies ensure that their political and policy goals are reflected in each such Summary for Policymakers (Henderson 2007a, 194).

C. Peer Review IPCC Style

It might be objected that whatever the points at which governments exert influence over the IPCC process, in the end, IPCC Assessment Reports are generated via a peer review process that is controlled by scientists, not politicians, and which therefore ensures that Assessment Reports reflect an objective evaluation of the existing scientific work on climate change. On the contrary, the IPCC peer review process seems only to enhance the unbridled discretion of chapter lead authors. How to organize IPCC peer review was not discussed until 1989, and "during the preparation of the First Assessment, the IPCC had no formal rules on who should participate in peer reviews, or formal follow up mechanisms. . . . " (Agrawala 1998, 624). By 1992, a formal task force on IPCC structure recommended that the process of selecting contributors and reviewers should be more open and that there should be clear guidelines for the time for reviews and peer review requirements for policymakers. Acting on these recommendations, the next year the IPCC established a two tier review system: the first by experts, and the second by all IPCC member governments and all experts and contributors from the first round. It was mandated that all expert review should include the "following categories: specialists with significant publications in particular areas, experts named in IPCC 'master lists' based on information supplied by various governments and organizations, and specialist reviewers nominated by international organizations including those in the UN system, the World Bank, Third World Academy of Sciences, OECD. . . . " (Agrawala 1998, 626).

It has been claimed that as these guidelines have been implemented, peer review of IPCC Assessment Reports is now more comprehensive, "by many orders of magnitude than that in the average journal. For example, draft chapters of the 1995 Working Group II Second Assessment report went through two full scale reviews: the first involving anywhere from twenty to sixty expert reviewers per chapter (a total of 700 experts from 58 countries were involved) . . . " (Agrawala 1998, 624). Similarly, in defending the IPCC process against charges of misconduct made after the release of the 1995 AR, Edwards and Schneider (2001, 237) describe one of the drafting Sessions for Working Group I—the climate science group—as an hours-long meeting involving six dozen climate scientists, many of whom were critics and were neither Lead Authors nor contributors, a session "remarkable" for its openness to diverse points of view, that added up to "peer review at ten times the normal level of scrutiny!" Thus according to some commentators, IPCC peer review is much more intense and critical than that found in peer-reviewed scientific journals.

This position, however, glosses over several important details of how the IPCC peer review process works. Up until 2000, the IPCC's practice in conducting reviews of its Assessment Reports was to make the coordinating Lead Authors and Lead Authors of Reports "responsible for evaluating and incorporating comments brought forward in the review" (Skodvin 2000, 410). This is equivalent to giving authors of papers submitted to peer-edited scientific journal the power to decide whether or not to pay attention to outside reviews in deciding whether their unrevised work merits publication. Unlike peer-reviewed scientific journals, where the decision whether to publish a paper, and if so, in what form, is decided by the editor, responding in varying degrees to outside reviewer opinions and comments, the IPCC process makes the Lead Author of an AR chapter effectively into the editor, and one that decides not whether or not to publish—for that decision has already been made—but whether to respond to outside comments.

There has been discussion within the IPCC about setting up editorial boards, similar to those used by scientific journals, but that has not yet been done (Agrawala 1998, 626). It is true that in 2000, the IPCC changed its outside review process by having the Bureau of each Working Group select two Review Editors from among scientists who were "not involved in the preparation or the review of the material for which they are an editor" (Skodvin 2000, 411). In conjunction with such Review Editors, Lead Authors are now required by IPCC procedures to "take account of expert and government review comments when revising text" (IPCC 2008, Annex 1, sections 4 and 5). Review Editors, however, are not like editors of scientific journals, with the ultimate authority about publication and content. Instead, their job is to ensure that Lead Authors do indeed "take account" of critical expert review by ensuring " . . . that all substantive expert and government review comments are afforded appropriate consideration. . . ," that " . . . genuine controversies are reflected adequately in the text of the Report. . . ," and, finally, that "where significant differences of opinion on scientific issues remain, such differences are described in an annex to the Report" (IPCC 2008 Annex 1, section 5). Moreover, whereas peer-edited scientific journals do not allow scientists submitting papers to choose either the editor or outside reviewers, under the IPCC procedures, the so-called Review Editors may themselves be members of an IPCC Bureau, and outside expert reviewers may be nominated by the very same Authors whose work they will be reviewing (IPCC 2008 Annex 1, sections 4 and 5). Hence as Agrawala (1998, 626) puts it,

> even though extensive peer review is conducted, the action on specific comments is left largely to the discretion of the writing teams. Technical Support Units do play a watchdog role but they can only persuade authors to follow up on review comments which writing teams may have inadequately dealt with. If these writing teams are dominated by opinionated experts holding one particular viewpoint, then, conceivably, they could get away with ignoring some or a majority of all critical review comments.

Another distinction between the IPCC expert review process and the peer-edited journal process is that under 2000 revisions in its rules of practice, the IPCC is free to consider and rely upon research that has not been published in peer-edited journals and in fact will likely never be published in such journals (Skodvin 2000, 414). This puts the Lead Authors of Assessment Report chapters in a position that is essentially that of an editor of a peer-edited journal, deciding which work is publishable and which is not. However, rather than waiting for submissions, as do peer-edited journals (aside from special issues with invited contributions), IPCC Lead Authors have the discretion to rely upon whatever unpublished sources they happen to know about and of which they approve. Hence, "[k]knowledge of and use of non-published material by IPCC Lead Authors will necessarily have to be arbitrary and network-related. This arrangement may . . . reinforce the notion of the IPCC as a 'club.'" (Skodvin 2000, 414).

A final procedural features makes it even more clear how the IPCC's version of "expert review" differs fundamentally from scientific peer review (see figure 2.2). This procedural feature mandates participation by developing country scientists and government actors on both IPCC Working Groups and on Lead Author teams (IPCC 2008, section 4.2.2). Early on in the history of the IPCC, the original conception of a "small core membership for each IPCC Working Group" was abandoned to "allow for widespread participation, particularly by developing countries" (Agrawala 1998, 628). In the words of Bert Bolin, the first IPCC Chair, "right now, many countries, especially developing countries, simply do not trust assessments in which their scientists and policymakers have not participated. Don't you think global credibility demands global representation?" (Agrawala 1998, 628). The problem identified by Bolin—of how to expand developing country participation—was a core concern of the IPCC, as shown by the fact that it was the only issue other than budget to be discussed at virtually every IPCC Bureau meeting from 1989 to 1996 (Agrawala 1998, 629). In 1988, there were only fourteen developing countries (out of a total of thirty countries) participating in the IPCC. By 1995, through a number of concrete steps, the IPCC had been transformed into an organization primarily made up of developing countries: of 117 countries participating in 1995, 98 were non-OECD members (Agrawala 1998, 630).

The steps to broaden developing country participation included what is essentially a quota requirement for developing country scientist participation on IPCC Working Groups and lead author teams. When new rules of IPCC procedure were adopted in 1993, "it was explicitly stipulated that for each chapter in the Second Assessment there should be at least one developing country lead-author [and that] . . . the chairmanship of each Working Group was to be shared by one developed and one developing country scientist."[7] These rules are no mere formality. They directly make a political consideration—of developing country representation in the IPCC process—trump scientific values. Working Groups and lead authors produce the Assessment Reports that are so grandly touted as objective assessments of the existing science. But rather than selecting the most scientifically qualified experts to serve on Working Groups and to write those reports (the lead authors), the IPCC has required that it select at least one member of each Working Group and each Lead Author from a developing country. While it may sometimes be easy to find a developing country scientist whose scientific qualifications would merit his or

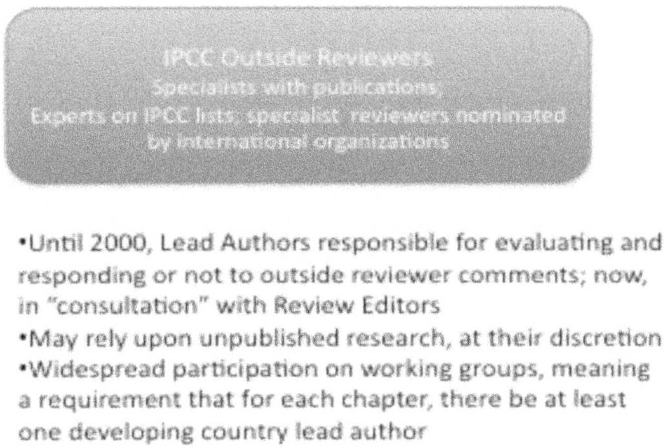

Figure 2.2. The Peer Review Process for IPCC Assessment Reports

her membership on a Working Group or Lead Author team even were she not from a developing country, the IPCC rules are such that if such a person cannot be found, then politics overrides science. It is hard to find any peer-edited scientific journal that relies upon such a nakedly political concern in determining editorial board membership.

D. Defending the IPCC Process: Views from the Climate Establishment

Climate scientists who have also been leaders in the IPCC have taken great pains to publicize the IPCC as an objective and disinterested evaluator of climate science, rather than as a science advocacy group. Consider, for example, how two of the leading IPCC physical scientists recently characterized that organization's activities in a recent lead editorial in the influential journal Science. After setting out the general view that as "the climate system continues to change and science continues to improve, so policy must be kept current with our best understanding" (Solomon and Manning 2008, 1457), leading IPCC scientists Susan Solomon and Martin Manning[8] go on to argue that the periodic IPCC Assessment Reports are crucial to the continuing usefulness of the IPCC. As they see the IPCC, its utility depends upon "its direct relevance to climate policy decisions" and this relevance requires periodic comprehensive IPCC assessments, because only these "span the breadth of policy-relevant issues and research advances, and they must be retained for a proper balance between science push and policy pull." In their view, the fact that the "IPCC does not plan or carry out research" is not a bad thing. Rather, it is not only consistent with the mandate of the IPCC "to evaluate information that must be independently documented, primarily as peer-reviewed literature," but "this separation between research and assessment is essential if the IPCC is to be an objective assessor." As Solomon and Manning describe it, the IPCC's "objective" Assessment Reports are ensured by a process of drafting, review and revision involving hundreds of authors and expert reviewers and dozens of governments. This process is time-consuming, but according to Solomon and Manning, "any move toward more rapid products risks incomplete identification of the range of justifiable views and a consequent reduction of the rigor, clarity and robustness of the consensus." According to Solomon and Manning, two key aspects of the IPCC process guarantee that it will be an "objective assessor": 1) the fact that IPCC does not plan or carry out research but instead "evaluates" information that is "independently documented" elsewhere; and, 2) the fact that IPCC Assessment Reports result from an "open" non-anonymous process in which authors, reviewers and governments are all identified by name.

A very similar but much more detailed defense of the IPCC process was published by Edwards and Schneider (2001). Their defense of the IPCC process was prompted by charges that in producing its Second Assessment Report, published in 1996, the IPCC had "corrupted the peer review process" for political reasons by allowing a chapter author (Benjamin Santer) to alter language in a key chapter in that report after it had been accepted at an IPCC plenary meeting. While a portion of Edwards and Schneider's defense relates to the particular charges made against the 1996 Report,[9] they also mounted a vigorous general defense of the IPCC process. Edwards and Schneider's defense stresses the "exceptional openness to critique, review and revision" of the entire process by which Working Group I produced the key Assessment Report dealing with climate science. As they describe it, such openness in the case of the 1996 Report included an hours-long meeting in Asheville, North Carolina, where chapter authors (Santer in particular) had to explain and defend their conclusions in front of dozen of scientists, heads of national weather bureaus, and officials from several countries. According to Edwards and Schneider (2001, 237):

> [t]his was peer review at ten times the normal level of scrutiny! It would be almost inconceivable for the editor of a peer-edited journal to duplicate this process. A few referees and an editor can only hope to execute the reviewing role a fraction as well as the remarkable open process at Asheville. Morever, after the Asheville meeting, two more IPCC drafts were written and reviewed by hundreds of additional scientists from all over the globe.

Edwards and Schneider acknowledge that the Asheville meeting and similar IPCC Working Group meetings were obviously far from "blind" review, so that "reputations, personalities, institutional politics, and the simple fatigue induced by long meetings probably played some role," but argue that "blind" review is only "one way to achieve the fundamental goals of peer review." In their view, "the quality improvements generated by the extensive and inclusive IPCC peer review process far outweigh the disadvantages of the open-meeting format for the final stages of peer review. If the real purposes of peer-review are . . . quality improvement, accountability, and community-building, then the IPCC process is as near to an ideal example as it may be possible to find." (Edwards and Schneider 2001, 238).

Of course, whether a process is "near to ideal" depends upon what one believes to be the purpose of the process. For Edwards and Schneider, the point of the IPCC process is clearly to communicate the "actual nature of the scientific consensus." They go on, importantly, to explain what they mean by "consensus":

> "Consensus" as we understand it, refers not to a single, exact prediction to which every scientist assents—an impossibility in this field—but to a generally agreed range of possible outcomes. This kind of consensus takes disagreement on details (and even, occasionally, on major points) for granted, both as an unavoidable element of a still-inexact science and as an important motor of scientific progress. Peer review, especially the inclusive and open form adopted by the IPCC, helps to build and maintain it. (Edwards and Schneider 2001, 244)

While somewhat opaque, this discussion makes clear that by "consensus" Edwards and Schneider do not mean the basic theories and facts that climate scientists take to be "true" descriptions of the earth's climate system. Indeed, they explicitly acknowledge that "this kind of scientific consensus is not the same thing as 'truth.' Once in a while, the skeptics are right" (Edwards and Schneider 2001, 245). Instead, it seems clear that Edwards and Schneider are not so much interested in what scientists know about the climate system, but about a "generally agreed" range of predictions about how the earth's climate will change as a result of human ghg emissions. But given that they acknowledge disagreements among climate scientists even on "major points," it seems that by "generally agreed" Edwards and Schneider must mean agreed upon by some subset of climate scientists. Given that they also endorse the open meeting process of "peer review" conducted by the IPCC, it seems that the subset of scientists whose agreement defines the "generally agreed . . . range of possible outcomes" are those whose views prevail at these IPCC meetings. Thus it seems that Edwards and Schneider, as well as Solomon and Manning (albeit more succinctly), conceive of the IPCC process as a way of adjudicating among claims and predictions on the frontier of climate science. In other words, in the eyes of the IPCC's defenders, "consensus" climate science means the scientific models, predictions and interpretations of observational evidence that the IPCC process ends up endorsing.

E. Supporting the Critique: The InterAcademy Council Report

As a response to the furor provoked by Climategate, in 2010 the IPCC and the U.N. asked an outside institution, the InterAcademy Council (IAC), to conduct a review of the IPCC's processes and procedures and to make recommendations for change if necessary. Funded by sponsors including national governments, private foundations and international organizations, the IAC is a "multinational organization of science academies" which produces reports on scientific, technical, and health issues at the request of international organizations (IAC 2010, ii). The IAC's Committee to Review the IPCC was comprised of nine academics, three representatives of government science agencies or national science organizations, and one representative of an intergovernmental agency that funds research in the biological sciences (the International Human Frontier Science Program Organization). In the narrow sense of not including any actual IPCC Panel members or scientists involved in writing IPCC Assessment Reports, the IAC Committee's composition avoided any conflict of interest. However, the majority of the Committee's members (such as scientist Mario Molina) had either expressly stated publicly their strong belief in the IPCC establishment view that anthropogenic CO_2 emissions could be singled out as the cause of recent climate change, or were leaders in organizations that were strong advocates of this position (Peter Williams of the UK Royal Society, for example), or had direct ties to the United Nations (such as Louise Fresco, who had been an Assistant Director the U.N.'s Food and Agriculture Organization). Aside from the pioneering but elderly climate modeler Syukuro Manabe, the Committee included no recognized leader in climate modeling.

Thus, based on its composition alone, one might well have expected that the IAC Committee's Report would have been strongly supportive of the IPCC's structure and process. This was decidedly not the case. It is true that in the IAC Report's Executive Summary concludes that the " . . . IPCC assessment process has been successful overall" (IAC 2010, 1). However, upon closer inspection, the IAC Report is very critical of the IPCC process. Most notably, the IAC Report pointed to the:

"[A]bsence of criteria for selecting key participants in the assessment process," such as the IPCC Chair, the Working Group co-chairs, members of the Bureau and the authors of the assessment reports (IAC 2010, 5);

Failure to follow the IPCC's own rules that unpublished or non-peer-reviewed sources are to be identified as such in the reference sections of Assessment Reports (IAC 2010, 19);

Failure to cite or discuss alternative views when Lead Authors do not agree with them (IAC 2010, 20);

The "near-universal" observation of all those contacted by the Committee that Review Editors do not have sufficient authority to ensure that chapter Authors consider "carefully" the comments of outside reviewers and document their responses (IAC 2010, 22);

Failure of the review process to be "truly independent because the Working Group Co-chairs, who have overall responsibility for the preparation of the reports, are also responsible for selecting Review Editors. To be independent, selection of Review Editors would have to be made by an individual or group that is not engaged in writing the report, and Review Editors would report directly to that individual or group" (IAC 2010, 23);

And, finally, the apparently insuperable and fundamental problem that "[d]espite the desirability of independent review, it is not clear what scientific body has the recognized legitimacy and capacity to carry out such a large task" (IAC 2010, 23).

Inasmuch as those who place great trust and confidence in the conclusions reached in IPCC Assessment Reports do so in large part because of what they perceive to be the "exhaustive" process of independent outside review called for by the IPCC's process, the IAC Report amounts to a devastating critique of the IPCC process. The IAC Report concludes, essentially, that the scientists who author chapters in IPCC Assessment Reports are free to completely disregard critical outside reviews unless a Review Editor who has been appointed by the same Working Group chairs who appointed the authors is able to persuade them to in some way acknowledge the outside critique. More simply, if the chapter authors, Review Editors and Working Group chairs all share the same views—as seems inevitable, given the total control vested in Working Group chairs—then for all practical purposes, the IPCC process is virtually designed to eliminate any serious consideration of competing scientific views.

III. THE COST OF IPCC ADVOCACY

In the previous part, I have attempted to persuade the reader that even if a description of the current "consensus" in climate science is indeed what policymakers want, it is not what the IPCC has provided. As we have seen, the IPCC has used its position as reporter on "consensus" to instead pick and choose among competing scientific claims on the frontier of climate science. In this part, I argue that IPCC advocacy has and will continue to have real costs: weakening the credibility of climate science, and failing to contribute to the formulation of wise policy toward anthropogenic ghg emissions.

A. Decentralization, Competition and the Credibility of Science

For many people, science consists simply of the definitions, facts, formulas and relationships that science textbooks set out as being true statements about the natural world.[10] Such a model of science seems to be what lawyers, policymakers and legislators mean when they say that laws and policies should simply be based on the "best available science"[11] or similar formulations: look at the newest textbook, and find the answer. For scientists themselves and philosophers of science, however, science is defined not by a set of facts and formulas currently believed to be true, but rather by a "method of investigation" or way of thinking.[12] Some philosophers have tried to come up with very narrow and precise definitions of the scientific method. Karl Popper, for example, defined the scientific method as the construction of hypotheses and the subsequent testing of those hypotheses against experience "by observation and experiment" (Popper 2002, 15). More precisely, on Popper's view, a system of hypotheses is scientific only insofar as it consists of universal statements or hypotheses that are capable of being "refuted by experience" (Popper 2002, 18–19).

The scientific method as conceptualized by Popper has come to be called the "hypothetico-deductive" method. For most scientists today, it is taken to be the "scientific method." As succinctly put recently by biologist Francisco Ayala (2009, 10035), the way that the scientific method works is as follows:

> New ideas in science are advanced in the form of conjectures or hypotheses, which may be more or less precisely formulated and be of lesser or greater generality. The tests to which scientific ideas are subjected include contrasting any hypothesis with the world of experience in a manner that must leave open the possibility that one might reject a particular hypothesis if it leads to wrong predictions about the world of experience. The possibility of empirical falsification of a hypothesis is carried out by ascertaining whether or not precise predictions derived as logical consequences from the hypothesis agree with the state of affairs found in the empirical world. A hypothesis that cannot be subject to the possibility of rejection by observation and experiment cannot be regarded as scientific.

Popper's empirical falsifiability criterion is an important elaboration of a more general view about science called logical positivism. On the positivist (and also on the Popperian) picture, scientists achieve consensus "based upon the dispassionate evaluation of objective empirical evidence . . . new contributions to science which were true and could be shown to fit the facts of nature were the ones which came to be accepted by the scientific community" (Cole 1992, 6). Unlike other areas of human endeavor, such as politics, on the positivist view,

> [t]he reaction of the scientific community to newly published work was believed to be determined by the content of that work . . . [and] the content of what could be accepted as fact was determined by nature. Scientists were trying to discover the laws of nature, which was a given and an external over which they had no control. If a paper put forth an idea which was not in accord with the laws of nature, attempts to replicate or support this idea would ultimately be doomed. (Cole 1992, 7)

For positivists, science steadily accumulates knowledge, with consensus the norm.

In his landmark 1962 book, *The Structure of Scientific Revolutions,* Thomas Khun attacked the positivist version of scientific method as fundamentally inconsistent with the actual history of science. According to Kuhn, scientific theories " . . . determine what evidence is looked for and what evidence is taken seriously" (Cole 1992, 7). Scientists cling to established theories even in the face of mounting disconfirming evidence, until a fundamentally new theory or "paradigm" is developed that better explains the existing and accumulating new observational data. Scientific developments such as the replacement of the Ptolemaic astronomical system by the Copernican system, or the acceptance of Newton's laws of motion cannot be described as cumulative accretions of knowledge, for entire conceptual structures were switched (Kuhn 2000, 14).

The importance of Kuhn's work has perhaps not so much been in his theory of scientific revolutions, for such revolutions are by definition rare, but rather in generating a body of research that questions whether the "hypothetico-deductive method"—the scientific method—itself suffices to actually explain the history of scientific change (Cole 1992, 15). One important strand of this work takes the history of science as precisely the process by which the "claims to truth" (Cole 1992, 15) of particular scientists working on the research frontier actually come to be generally accepted as both true and important and therefore part of the core or consensus knowledge of the discipline.

While some sociologists believe that the content of a new scientific paper has nothing to do with whether it is accepted into the "core" of the discipline, that acceptance is solely a function of sociological factors (see, e.g., Latour 1987), a more nuanced position, seemingly more in accord with the existing empirical evidence, is that while sociological factors matter, so too does the actual truth or falsity of a particular scientific claim (Cole 1992). As Bauer (2004, 646) describes this process:

> Science begins as hunches. Those that work out become pieces of frontier science. If competent peers think it worthy of attention, an item gets published in the primary research literature. If other researchers find it useful and accurate, eventually the knowledge gets into review articles and monographs and finally textbooks. The history of science demonstrates that, sooner or later, most frontier science turns out to

> need modifying or to have been misleading or even entirely wrong. Science employs a knowledge filter that slowly separates the wheat from the chaff.

From an economic point of view, the scientific "knowledge filter" may be thought of as the process by which scientific claims are certified as true and important. Because of the tremendous asymmetry in knowledge between users and creators of new scientific knowledge, certification in science takes on a "massively important" role (Turner 2002, 375). There are, correspondingly, multiple certifying institutions in science: academic institutions, which decide whether to hire and promote scientists; governmental and other funding bodies, who must decide who and what to fund; professional organizations that may bestow prizes for exceptionally important research; and the editors of peer-edited journals, who decide whether scientific claims are fit for publication.

Of course, none of the institutions that certify science are perfect. Consider, for example, peer review prior to publication. The requirement of peer review prior to publication seems designed to act to screen the likely truth and significance of work that is published. Such initial screening reduces the probability that scientists waste their time attempting to extend or build upon false positives—false claims to truth—or trivially true claims. However, peer review is certainly not perfect in weeding out only significant new claims, as lots of articles are never cited after publication (see van Dalen and Klamer, 2005, discussing data for economics journals, and Cole 1970 discussing physics journal articles). Nor does peer review always catch and screen out fraudulent or false claims (see, e.g. Shafer 2009, discussing a notorious case of fifteen years' worth of fraudulent publications by a particular researcher dealing with the benefits of peri-operative non-steroidal anti-inflammatory drug administration). Conversely, as shown by the difficulty that future highly cited and even Nobel-prize-winning work often has had in initially being published (Campanario 1996, 2009), peer review editors and referees sometimes are so skeptical about new claims to truth that publication in a peer-reviewed journal becomes a real hurdle for work that is not only true but revolutionary.

Importantly, publication in a peer-reviewed scientific journal is not the end of the scientific certification process. As observed by Turner (2002, 377), because the publication of a new, testable hypothesis "invariably" has implications for previous hypotheses, in a very real sense, all science involves both production and authentication or assessment. As other scientists attempt to build upon and extend earlier published results, they are testing those results. In addition, as seems increasingly to be true, early work cannot be continued unless grant-making bodies determine that continuing down a particular research avenue is likely worthwhile. The vetting of claims to scientific truth also takes place in the academic promotion process, and in the various activities of professional scientific associations, such as prize competitions and leadership determinations.

It is this congeries of certification institutions that constitutes the scientific process. What is vital to the credibility of this process is that there is no single scientist or institutional body that is authoritative. At any given time, different individual scientists, university departments, and grant-making institutions may disagree about the merit of particular line of research. Such disagreement is not a bad thing: for as Turner (2002) has argued, it is precisely competition among certifiers—different scientists, journals, academic departments and funding institutions—that ensures the credibility of certification. Competition ensures that if a group of scientists or even an entire professional association attempts to discredit important and true new findings, simply because they threaten the established view maintained by the entrenched group, then they will fail. Some other journals, not edited by members of the entrenched group, will publish the new work; departments at other universities will hire and promote the pathbreaking scientists; alternative funding organizations will support continued research stimulated by the new findings.

The reason why a decentralized and competitive certification process is so vital to the credibility of science is precisely because the scientific process relies so heavily on the opinions of scientists themselves. Certification of science by scientists ensures that experts are reviewing new work, thus overcoming the general lack of such expertise among users of science. But certification by scientists working in the very same field raises inherent conflict of interest problems. Scientists and academic departments have an interest in pushing the publication and dissemination of work that recognizes and builds upon their own work while suppressing or discrediting conflicting work; journal editors may wish to further their own citation rates while discouraging citation of work in competing journals; funding agencies likewise have an interest in funding further work that tends to confirm the correctness of their earlier funding decisions while denying funds to researchers whose proposals threaten to show that earlier-funded work was incorrect or unimportant. As Turner (2002) suggests, the well-known norms of science publicized by Merton (1973) may indeed all be understood as a response to the demand by users of science that certifiers be free from self-interested bias: objectivity and freedom from known bias are vital to a journal's value as a certification institution; "disinterestedness" is likewise valued as the opposite of bias; "organized skepticism" describes a market in which certifying or bonding institutions compete for credibility; finally, scientific "universalism" is the value produced by a decentralized, competitive market for scientific certification. As Turner (2002, 381) so aptly puts it, "[i]f we can trust the market for bonding to produce adequate bonders, we can trust the results of science where there are such markets . . . it is the market for bonding in which we place our trust, and on which our trust in science properly rests."

B. Attempted Monopolization in the Science Certification Market: The IPCC's Harmful Effect on Climate Science Credibility

In light of this understanding of the scientific process as a competitive, decentralized market for the certification of claims to scientific truth, the IPCC's structure, process and overall role in the production and certification of climate science is deeply troubling. The first and most severe problem is that the IPCC has interpreted its task as supplying a policy-relevant determination of the truth of competing claims on the frontier of climate science, a determination that is premature in that it short-circuits the normal, decentralized process of scientific truth certification. Examples of this abound, but several of the more striking in the IPCC Assessment Reports include: a virtually complete failure to discuss and acknowledge the significance of a well-established body of work showing that black carbon (soot)—rather than CO_2—is responsible for the bulk of Arctic warming over the last 100 years; the failure to clearly acknowledge the substantial uncertainty over the size and sign of crucial feedback effects that entirely determine projected future temperature increases due to CO_2 increases; publicizing as uncontroversial and "likely" projections of species loss due to climate change that rely upon a new methodology that a very large number of biologists have rejected as invalid in peer-edited articles; and, finally, dismissal of competing explanations for late-twentieth-century warming—natural climate cycles and solar variation—that remain contested in the peer-reviewed literature. These are but a few of the more striking examples, but they all show the IPCC intervening to judge the winning, true side in scientific disputes that are quite new and as yet undecided by the traditional vetting institutions of science. The IPCC is essentially trying to establish itself as the authoritative, monopoly provider of climate science certification.

Even worse, the Climategate emails reveal lead IPCC scientists collectively intervening in the traditional scientific process in an attempt to stifle and prevent competition. From an economic or business point of view, efforts by groups of like-minded scientists to pressure journal editors to delay the publication of disconfirming work and to speed the publication of supportive work, and to boycott non-complying journals entirely—as revealed, for example, by the Douglass et al. v. Santer et al. dispute over tropical tropospheric warming—are classic attempts by one group of competitors to stifle competition by another group. Now in a competitive market, such efforts usually fail and cause little harm: no group of competitors has market power that is destined to last. However, in the case of group efforts by lead IPCC climate scientists, such group efforts are likely to cause very real harm in the market for climate science certification. The reason they are likely to be harmful is because these group efforts by lead IPCC scientists are an attempt to leverage the monopoly power that the IPCC is attempting to establish in the climate science certification market over to the otherwise decentralized and competitive world of climate science journals. To be blunt, when the scientists who run the IPCC process tell a journal editor that they want publication of a competing paper delayed and of their own paper speeded up, they are implicitly making a very real threat: that if the editor fails to comply, that failure will be remembered and punished in future decisions not only about where to publish, but in IPCC decisions about which climate science to certify as true and important.

The deleterious impact of the IPCC on the world of peer-edited climate science journals extends beyond collective efforts to influence the supposedly independent decisions of journal editors, to impact the content of what counts as "climate science." As Roger Pielke Sr. has noted (2009), it has become commonplace in several peer-edited climate journals to publish computer projections of far-distant climate variables. The papers of course cannot test far-distant projections against any data. Thus at the best, these papers set out as-yet untested hypotheses; at the worst, they set out hypotheses which are in principle untestable. Thus the crucial step in the scientific process—empirical testing—is increasingly simply not required by many climate science journals. Rather, as Lindzen (2008) has noted, the climate science literature increasingly has come to speak of computer model projections not as something that needs to be tested against observational data, but as themselves constituting "evidence." Were model projections alone taken as "evidence," then all kinds of false theories would be confirmed simply by building a computer model to spit out future projections generated by the theories: indeed, under such a practice, computer models of future macro-economic behavior, which have generally been completely unable to accurately predict even short run macroeconomic variables, would have to be taken as somehow confirmed by the evidence. Most fundamentally, the practice of treating computer model projections as evidence is flatly inconsistent with a scientific process designed to provide credible certification of true scientific claims. But it is precisely what one would expect in a system where climate science journals understand their mission as providing what the IPCC demands: the imprimateur of publication in peer-reviewed journals attached to projections of far-distant future climate that the IPCC wishes to then authoritatively and monopolistically certify as true.

As for the funding of climate science, the IPCC's influence has been so great that it has likely effectively obliterated most of the evidence of its influence. The IPCC leadership is virtually isomorphic to the leadership of the federal agencies that fund climate science in the United States, such as NOAA Climate Science, NASA and the National Science Foundation.

For this reason, one would rationally expect a very low probability that any of these agencies will actually fund proposed scientific work that challenges any of the core premises in the establishment climate science story. Given such a low probability of funding, it would be foolish for a climate scientist to spend any serious time or effort on making such a proposal to these agencies. Hence in the universe of submitted proposals to climate science funding agencies, one would expect to find few that even proposed work questioning the IPCC's preferred view of climate science; if the primary effect of the IPCC's monopoly power is to deter dissenting scientists from even attempting to get public funding, then that power would have an enormous impact on climate science, but be empirically difficult to identify.

A few dissenting climate scientists have, however, submitted proposals for funding from the relevant federal science agencies. Their experience, while limited, suggests that funding decisions are being driven by a highly centralized view of what constitutes significant and fundable new work in climate science, as opposed to the more decentralized opinions of climate scientists serving as outside referees. Pielke Sr. (2010), for example, provides evidence of uniformly very positive outside referee reports on proposals for NSF funding submitted by him and his colleagues, but persistent funding denials by the relevant NSF Program Officer. Gavin Schmidt (2010), a lead IPCC scientist, has provided a "candid" description of the grant review process. His remarks confirm the image of a grant-making process that would strongly transmit the preferred views of members of a monopolistic certification institution. Schmidt extols the apolitical nature of the process, but freely admits that having a grant review panel member with expertise on the proposed topic made the proposal more likely to be either quickly approved or dismissed, and that "[t]he reputation of the proposers as people of good science goes a long way to judging the feasibility of the proposal." The process he describes is almost perfectly designed to transmit the influence of the IPCC: by certifying expertise, the IPCC influences the composition of grantmaking panels, whose members can then ensure that funding goes only to work which is consistent with the IPCC view of what is true and important in climate science.

In addition to representing the intrusion of centralized, monopolistic control over the climate science certification process, the IPCC has also and perhaps more obviously made policy preferences a key factor in that process. No one denies that politics is inherently part of the IPCC process: the Panel itself after all consists of government representatives, and it is the Panel that chooses the Bureau, which then chooses the Working Group chairs, who in turn choose the Chapter Authors and Review Editors who determine the content of IPCC Reports. Alternative answers to the key questions on the frontier of climate science, such as the determination of the sign and magnitude of feedback effects, or the role of black carbon in Arctic warming, all have known policy consequences. If feedback effects are large and positive, then the benefits of ghg emission reduction are much bigger than if some of those feedback effects are negative. If the large impact of black carbon on Arctic warming is factored in, then the benefits of reducing human ghg emissions are likely much smaller than if black carbon is ignored. Thus if one knows where a particular scientist stands on these key questions, one knows the policy implications of that scientist's work; correspondingly, any scientist who has policy preferences with respect to human ghg emissions knows how the scientific questions need to be answered to support those policy preferences. Given the direct relationship between scientific answers and policy prescriptions in the global warming area, the IPCC can produce "scientific" reports that are guided by implied policy consequences. Moreover, given this direct relationship between science and policy, the remarkable consistency of the pattern of intervention by IPCC scientists in the climate science process—advocating scientific work that supports the IPCC political agenda of quick action to reduce human ghg emissions, and suppressing and discrediting work that weakens the case for immediate drastic emission reductions—seems strong evidence of a politically driven scientific agenda. To put the point as a question: if IPCC scientists were intervening in the scientific process solely in the support of the search for true hypotheses, wouldn't there be at least one case where they intervened to support a hypothesis that tended to weaken the case for immediate reductions in human ghg emissions?

C. The Impact of IPCC Science Advocacy on the Policy Process

The IPCC's science agenda is clear: to make the case for the primacy of human ghg emissions—primarily CO_2—in causing global warming, to stress that the evidence in favor of this CO_2 primacy hypothesis has gotten stronger and stronger over the past couple of decades, and to proclaim the likelihood of all kinds of catastrophic harm from accelerated future warming if anthropogenic ghg emissions are not reduced. If one believes what the IPCC says, then the clear policy consequence is that provided that one does not discount the interests of future generations too heavily, almost any amount of money spent in reducing present-day human ghg emissions is potentially cost-benefit justified. The policy implications are blunt: almost any policy that effectively reduces present and future ghg emissions is, on this view, justified. For example, crude policy responses such as banning the use of coal as a fuel for electric power generation and massively subsidizing the construction of new wind and solar power facilities and a new power grid across America would all be justified. The only reason why someone who believes what the IPCC says would proceed more cautiously—endorsing, for example, a national program with a declining cap on CO_2 emissions but allowing individual firms to meet their caps by buying permits from other firms—is the constraint of political feasibility.

Suppose that instead of acting as an advocate for these positions, the IPCC had instead produced and published a Report with the express purpose of fully informing the public about all of the work on the climate science frontier, both that which supports the CO_2 primacy hypothesis and that which does not, and about all of the open and highly policy-relevant questions in climate science. Such a Report, which would have been an unbiased and objective assessment rather than brief in favor of a particular position, would have had at the very least included a much more open and informative discussion of not only the degree of uncertainty in climate model projections but the causes of such uncertainty, the basic fact of incomplete knowledge of many key processes (compare Tol 2011). Moreover, such an unbiased and objective appraisal would have had drastically different policy implications than have the IPCC's Assessment Reports. These implications would have included the following:

1) Warming in the Arctic region in particular could likely be significantly ameliorated if China and other countries with rapidly growing fossil-fuel-based industrial sectors adopt traditional air pollution reduction technologies;

2) Given the very real positive probability of only small future increases in temperature even if global CO_2 emissions continue to increase (due to small positive feedback effects), policies to reduce CO_2 emissions are likely to be cost-benefit justified only if they are easily reversible and if they bring some benefit other than reducing the risk of future harm from global warming;

3) As it is almost surely the case that some regions of the world will benefit from global warming in many possible temperature ranges, costly ghg emission reduction is not a global public good, but instead an economic good in some regions and an economic bad in others; given this, an effective global ghg emission reduction treaty through the auspices of the U.N. is extremely unlikely—effective international agreements to reduce ghg emissions require actual exchanges and compensation and are much more likely at the bilateral or regional level;

4) As we learn more about natural cycles in global climate, approaches to human infrastructure development that anticipate cycles, rather than assuming a constant or steady-state climate, ought to be pursued.

IV. TOWARD THE FUTURE: CLIMATE SCIENCE WITHOUT THE IPCC

Perhaps the most significant finding of the 2010 IAC Report on the IPCC is that the conflict of interest that lies at the heart of the process by which IPCC Assessment Reports are produced—the fact that Working Group Chairs choose Lead Authors and also the Review Editors who are supposed to monitor Lead Authors—cannot be fixed within the IPCC itself. The simple reason is that there is nobody within the IPCC structure who is independent of the political agenda of the IPCC, and therefore nobody within the IPCC who would be capable of independent selection of IPCC Review Editors. As the IAC Report suggests, to fix the IPCC would require the existence of some other, more objective and less biased institution which could credibly choose Review Editors, and also probably chapter Lead Authors. This other institution would presumably have to be relatively apolitical, in the sense of having no particular stake in a particular policy response to global warming, and also scientifically eclectic, in that it would somehow ensure the participation of scientists from both sides of ongoing disputes on the frontier of climate science.

Such an institution does not now exist and it is far from clear how it could ever come into existence. Even if such an institution did exist, there would remain a more fundamental problem with the IPCC: its mission. Since its inception, the mission of the IPCC has been to answer a single overarching question: whether the scientific evidence supporting the view that human ghg emissions may cause serious harm is sufficiently strong that costly international action should be taken to reduce such emissions. This question is itself leading—for how could the IPCC ever conclude that the evidence in favor of dangerous human ghg emissions will never be strong enough to support costly emission reductions? It also puts the cart before the horse, by asking not about the state of climate science, but whether climate science can yet support a particular policy. Given that the IPCC's mission was to answer a highly leading, policy-driven question put by its creators and supporters, it seems unrealistic to think that the IPCC could simply have continued, in one Report after another, to provide an answer of the form "climate science is not yet developed enough to provide an answer to the question asked." If supporters of the IPCC hadn't eventually gotten the answer that they wanted, then the IPCC would have been replaced.

The IPCC is thus an institution whose mission guaranteed that it would eventually provide a pre-determined answer. As argued above, it has provided an answer that is so uninformative about the true state of climate science that it has a very high probability of supporting the wrong policy responses; even worse, the IPCC has supported the cartelization of climate science in the service of climate policy, likely retarding the progress of climate science and certainly harming its credibility. With such a flawed mission and such harmful consequences, the IPCC must be abolished.

Abolishing the IPCC would not eliminate the need for policy-relevant advice about the current state of climate science. In fashioning an institution for providing such advice, there would remain the fundamental problem, discussed in the chapter by Conrad, that in any area of rapidly evolving scientific knowledge, some of the scientific answers that policymakers are seeking simply do not yet exist. It may be that the appropriate institutional design would reflect the kind of adversarial structure advocated by McKitrick in his chapter. My own view is that a more definitive answer to the question of how to design institutions for the assessment of the state of policy-relevant science ought to turn on further research into the incentives created by alternative institutional designs. The one thing that does seem clear is that the IPCC's structure is one to avoid.

NOTES

1. Helpful comments on this material were provided by participants in a University of Pennsylvania law faculty retreat and by Cary Coglianese, P. David Henderson, and Roger Pielke Sr.

2. As quoted in Henderson (2007a, 195).

3. As observed by Agrawala (1998, 633), the IPCC's policymaker summaries have had a direct impact on policy: "the summary produced by Working Group I for its First Assessment is widely regarded as being very authoritative. . . . its statement that (under a business as usual scenario) the world is likely to see 'a rate of increase of global mean temperature during the next century . . . that is greater than that seen over the past 10,000 years' was, by most accounts, very influential in catalyzing the decisionmaking process which eventually led to the signing of the FCCC in 1992."

4. As I have noted elsewhere, (Johnston 2010), subsequent work on the relationship between predicted and observed tropical tropospheric temperature trends has tended to confirm Douglass et al. (2007) in finding the observed increases are less than models predict. Also and more importantly, that work (see Bengtsson and Hodges 2009; McKitrick, McIntyre, and Herman, 2010) has quite clearly demonstrated that the statistical approaches taken by both Douglass et al. (2007) as well as Santer et al. (2008) are incorrect.

5. This would seem to violate norms in the economics profession. See List et al. (2001).

6. Much of my description follows that presented by David Henderson (2007a).

7. In addition, the IPCC now requires that it pay for at least one developing country expert to attend each writing team meeting of very IPCC chapter (Agrawala 1998, 630).

8. According to the capsule biographies accompanying the Science editorial under discussion, Susan Solomon is co-chair of IPCC Working Group I and a senior scientist at the Earth System Research Laboratory of NOAA, while Martin Manning, a professor at Victoria University in New Zealand, was formerly the head of the IPCC Working Group I Technical Support Unit.

9. They argue that what the chapter author Benjamin Santer did in altering Report language after acceptance by the IPCC Plenary was not inconsistent with what IPCC procedures then allowed. It is worth noting, however, that the IPCC's procedures were later revised to prohibit such post-acceptance alteration of Report language. See IPCC (2008, section 4.2).

10. Paraphrasing loosely Derry (2002, 3).

11. Congress has been especially fond of this particular phrase, found in important federal environmental statutes such as section 4 of the Endangered Species Act, which sets out the standard for listing of species as endangered or threatened.

12. Derry, "What Science is and how it Works."

REFERENCES

Agrawala, Shardul (1998). "Structural and Process History of the Intergovernmental Panel on Climate Change." *Climatic Change* 39: 621–42 .

Ayala, Francisco J. (2009) "Darwin and the Scientific Method." *Proc. Nat'l Acad. Sci.* 106: 10033–10039.

Baker, Marcia B., and Gerard H. Roe (2009). "The Shape of Things to Come: Why is Climate Change so Predictable?" *Journal of Climate* 22: 4574–4589.

Bauer, Henry H. (2004). "Science in the 21st Century: Knowledge Monopolies and Research Cartels." *Journal of Scientific Exploration* 18(4): 643–60.

Bengtsoon, Lennart, and Kevin L. Hodges (2009). "On the Evaluation of Temperature Trends in the Tropical Troposphere." *Climate Dynamics* doi 10.1007/S00382-009-0680.

Biello, David (2007). "Climate Change Verdict: Science Debate Ends, Solution Debate Begins: the IPCC Summary for Policymakers Definitively Proclaimed the Globe to be Warming as a Result of Human Activity, Now the Science Shifts to Impacts and Solutions." *Scientific American* February 8, 2007.

Boehmer-Christiansen, Sonja (1994). "Global Climate Protection Policy: the Limits of Scientific Advice," Part 1. *Global Envl. Change* 4(2): 140–59.

Botkin, Daniel B., Henrik Saxe, Miguel B. Araujo, Richard Betts, Richard H. W. Bradshaw, Thomas Cedhagen, Peter Chesson, Terry P. Dawson, Julie Ro Etterson, Daniel P. Faith, Simon Ferrier, Antoine Guisan, Anja Skjoldborg Hansen, David W. Hilbert, Craig Loehle, Chris Margules, Mark New, Matthew J. Sobel, and David R. B. Stockwell (2007). "Forecasting the Effects of Global Warming on Biodiversity." *Bioscience* 57(3): 227–36.

Campanario, Juan Miguel (2009). "Rejecting and resisting Nobel Class Discoveries: Accounts by Nobel Laureates." *Scientometrics* 81(2): 549-65.

Campanario, Juan Miguel (1996). "Have Referees Rejected Some of the Most Cited Papers of All Times?" *J. American Society Information Sciences* 47(4): 302–10.

Cole, Stephen (1992). *Making Science: Between Nature and Society*. Cambridge, MA: Harvard University Press.

Cole, Stephen (1970). "Professional Standing and the Reception of Scientific Discoveries." *Amer. Soc. Rev.* 76(2): 286–306.

Derry, Gregory N. (2002). *What Science Is and How It Works*. Princeton, NJ: Princeton University Press.

Douglass, D. H., and J. R. Christy (2009). "A Climatology Conspiracy?" *American Thinker*, December 20, 2009, available at www.americanthinker.com/2009/12/a_climatology_conspiracy.html.

Douglass, D. H., J. R. Christy, B. D. Pearson, and S. F. Singer (2007). "A Comparison of Tropical Temperature Trends with Model Predictions," *International Journal of Climatology* 27: doi: 10.1002/joc.1651.

Edwards, Paul N., and Stephen H. Schneider (2001). "Self-Governance and Peer Review in Science-for-Policy: The Case of the IPCC Second Assessment Report," in *Changing the Atmosphere: Expert Knowledge and Environmental Governance*, edited by Clark A. Miller and Paul N. Edwards, 219–46. Cambridge, MA: MIT Press.

Farber, Daniel A. (2007). "Adapting to Climate Change: Who Should Pay?" *J. Land Use* 23: 1–37.

Flanner, Mark G., Charles S. Zender, James T. Randerson, and Philip J. Rasch (2007). "Present-Day Climate Forcing and Response from Black Carbon in Snow." *Journal of Geophysical Research* D11202, doi:10.1029/2006JD008003.

Henderson, David (2007a). "Governments and Climate Change Issues: The Case for Rethinking." *World Economics* 8(2): 183–228.

Henderson, David (2007b). "Unwarranted Trust: A Critique of the IPCC." *Energy & Environment* 18(7–8): 909–28.

Holland, David (2007). "Bias and Concealment in the IPCC Process: The 'Hockey-Stick' Affair and its Implications." *Energy & Environment* 18(7–8): 951–83.

Intergovernmental Panel on Climate Change (IPCC) (2011). "Principles Governing IPCC Work Approved at the 14th Session (Vienna, 1–3 October, 1988), amended at the 21st Session (Vienna, 3 and 6–7 November, 2003) and at the 25th Session (Mauritius, 26–28 April 2006)," available at www.ipcc.ch/about/princ.pdf.

IPCC (2008). "Procedures for the Preparation, Review, Acceptance, Adoption, Approval and Publication of IPCC Reports, Appendix A to Principles Governing IPCC Work." Available at www.ipcc.ch/pdf/ipcc-principles/ipcc-principles-appendix-a.pdf.

IPCC (2007). "Climate Change 2007: The Physical Science Basis," edited by Susan Solomon, Dahe Qin, Martin Manning. Cambridge: Cambridge University Press.

IPCC (2007a). "Climate Change 2007: Climate Change Impacts, Adaptation, and Vulnerability," edited by Martin L. Parry, Osvaldo Canziani, Jean Palutikof, Paul van der Linden, and Clair Hanson. Cambridge: Cambridge University Press.

Johnston, Jason Scott (2010). "Global Warming Advocacy Science: A Cross Examination," available at lsr.nellco.org/upenn_wps/324/.

Khandekar, Madhav (2009). "Global Warming and Sea Level Rise," *Energy & Environment* 20: 1067–1074.

Kuhn, Thomas (2000). "What are Scientific Revolutions?" In Thomas Kuhn, *The Road Since Structure: Philosophical Essays 1970–1993*. Chicago: University of Chicago Press.

Landsea, Christopher (2005). "Letter to the Community," available at sciencepolicy.colorado.edu/prometheus/archives/science_policy_general/000318chris_landsea_leaves.html.

Latour, Bruno (1988). *Science in Action*. Cambridge, MA: Harvard University Press.

Lewis, Owen T. (2006). "Climate Change, Species-Area Curves and the Extinction Crisis," *Philosophical Transactions of the Royal Society Series B* 361: 163–70.

Lindzen, Richard (2008). "Climate Science: Is it Currently Designed to Answer Questions?" available at climatephysics.com/2008/08/31/lindzen-climate-science-is-it-currently-designed-to-answer-questions/.

List, John A., Charles D. Bailey, Patricia J. Euzent, and Thomas L. Martin (2001). "Academic Economists Behaving Badly? A Survey on Three Areas of Unethical Behavior," *Economic Inquiry* 39: 162–70.

McKitrick, Ross, Stephen McIntyre, and Chad Herman (2010). "Panel and Multivariate Methods for Tests of Trend Equivalence in Climate Data Series," *Atmospheric Science Letters* 11: 270–77.

McIntyre, Stephen, and McKitrick, Ross (2005). "Hockey Sticks, Principal Components, and Spurious Significance," *Geophysical Research Letters* L03710, doi: 10.1029/2004GL021750.

Merton, Robert K (1973). "The Normative Structure of Science," in *The Sociology of Science: Theoretical and Empirical Investigations*. Chicago: University of Chicago Press.

Pielke Sr., Roger, (2010). "Is the NSF Funding Process Working Correctly?" Available at pielkeclimatesci.wordpress.com/2010/05/18/is-the-nsf-funding-process-working-correctly.html.

Pielke Sr., Roger (2009). "Short Circuiting the Scientific Process—A Serious Problem in the Climate Science Community," available at pielkeclimatesci.wordpress.com/2009/short-circuiting-the-scientific-process-a-serious-problem-in-the-climate-science-community.html.

Pielke Sr., Roger, Christopher A. Davey, Dev Niyogi, Souleymane Fall, Jesse Steinweg-Woods, Ken Hubbard, Xiaomao Lin, Ming Cai, Young-Kwon Lim, Hong Li, John Nielsen-Gammon, Kevin Gallo, Robert Hale, Rezaul Mahmood, Stuart Foster, Richard T. McNider, and Peter Blanken (2007). "Unresolved Issues with the Assessment of Multidecadal Global Land Surface Temperature Trends," *Journal of Geophysical Research* D24S08, doi: 10.1029/2006JD008229.

Pielke, Jr., Roger A. (2006). "When Scientists Politicize Science," *Regulation* 29(1): 28–34.

Popper, Karl (2002). *The Logic of Scientific Discovery*. New York: Routledge.

Roe, Gerard H., and Marcia B. Baker (2007). "Why is Climate Sensitivity so Unpredictable?" *Science* 318: 629–33.
Rosenthal, Elisabeth (2007). "U.N. Report Describes Risks of Inaction on Climate Change," *New York Times*, November 17, 2007.
Santer, B. D. (2010). Open Letter to the Climate Science Community: Response to "A Climatology Conspiracy?" Available at www.realclimate.org/docs/open_letter_3_to_Community.pdf.
Santer, B. D., P. W. Thorne, L. Haimberger, K. E. Taylor, T. M. L. Wigley, J. R. Lanzante, S. Solomon, M. Free, P. J. Gleckler, P. D. Jones, T. R. Karl, S. A. Klein, C. Mears, D. Nychka, G. A. Schmidt, S. C. Sherwood, and F. J. Wentz (2008). "Consistency of Modeled and Observed Temperature Trends in the Tropical Troposphere," *Int'l J. Climatology* 28: 1703–1722. doi: 10.1002/joc.1756.
Scafetta, Nicola, and West, Bruce J. (2008). "Is Climate Sensitive to Solar Variability?" *Physics Today*, March 2008, 50–51.
Schmidt, Gavin (2011). "Reflections on Funding Panels." Available at www.realclimate.org/index.php/archives/2011/01/reflections-on-funding-panels/#more-4972.
Schwartz, Stephen E, Robert J. Charlson, and Henning Rodhe (2007). "Quantifying Climate Change—Too Rosy a Picture?" Nature Reports: Climate Change 2:23–29.
Shafer, S. (2009). "Tattered Threads," *Anesth. Anal.* 108: 1361–1363.
Skodvin, Tora (2000). "Revised Rules of Procedure for the IPCC Process: An Editorial Essay," *Climatic Change* 46(4): 409–15.
Solomon, Susan, and Martin Manning (2008). "The IPCC Must Maintain its Rigor," *Science* 319: 1457.
Stott, Lowell, et al. (2007) "Southern Hemisphere and Deep-Sea Warming Led Deglacial Atmospheric CO_2 Rise and Tropical Warming," *Science* 318: 435–39.
Sun, De-Zheng, Yongqiang Yu, and Tao Zhang (2009). "Tropical Water Vapor and Cloud Feedbacks in Climate Models: A Further Assessment Using Coupled Simulations," *Journal of Climate* 22: 1287–1304.
Tol, Richard S. J. (2011). "Regulating Knowledge Monopolies: the Case of the IPCC," *Climatic Change* 108: 827–39.
Turner, Stephen (2002). "Scientists as Agents," in *Science Bought and Sold*, 362–84, edited by Philip Murkowski and Esther-Mirjam Sent. Chicago: The University of Chicago Press.
Van Dalen, Hendrik P., and Arjo Klamer (2005). "Is Science a Case of Wasteful Competition?" *Kyklos* 58: 395–414.
Wang, J.-W., K. Wang, R. A. Pielke, J. C. Lin, and T. Matsui (2008). "Towards a Robust Test on North American Warming Trend and Precipitable Water Content Increase," *Geophysical Research Letters* 35: L18804, doi:10.1029/2008GL034564.
Wunsch, Carl, Rui M. Ponte, and Patrick Heimbach (2007). "Decadal Trends in Sea Level Patterns: 1993–2004," *Journal of Climate* 20: 5889-5911.
Zillman, John W. (2007). "Some Observations on the IPCC Assessment Process 1988–2007," *Energy & Environment* 18(7–8): 869–92.

THREE

Adversarial versus Consensus Processes for Assessing Scientific Evidence

Should the IPCC Operate More Like a Courtroom?

Ross McKitrick

I. INTRODUCTION

Our government uses two contrasting systems to adjudicate complex questions upon which significant consequences depend. In criminal or civil court actions an adversarial system is used, in which openly biased representatives advocate on behalf of two opposing sides in front of a presumably neutral third party. In the case of scientific assessments, such as those undertaken by the Intergovernmental Panel on Climate Change (IPCC), adversarial subgroups are not recruited, instead a single panel is formed in which members try to be neutral and collaborative for the purpose of reaching a consensus. Such collaboration would be viewed as a violation of proper procedures in a courtroom: if a defense attorney were found to be assisting the prosecution prepare its case, that would be considered a form of corruption. By contrast, such collegial interaction would be considered constructive and beneficial in an expert panel. Indeed if a potential panel member were known to be committed to one particular point of view, it would potentially disqualify him or her from serving; yet such a commitment is an essential feature of being an effective advocate in court.

If we were to envision a court-like adversarial structure for a scientific assessment panel, it would involve defining three distinct groups: one to argue for a certain proposition, one to argue against, and one to supervise the process (as well as potentially to deliver a verdict). It is easily supposed that a scientific topic might have identifiable proponents of conflicting points of view, since a question of sufficient complexity as to require an expert assessment is almost certain to be subject to ongoing uncertainty and disagreement, and experienced researchers in the field will no doubt be aware of scholars associated with different perspectives. This is definitely true in the case of climate change, which is the particular example of interest in this chapter.

If it is known that there are contrasting opinions on a particular issue, the structure of a court trial provides an obvious framework for ensuring each side is represented and each side's evidence is subject to adversarial examination. But the expert panel structure does not necessarily provide similar guarantees. While rules may be in place to try and ensure both full representation of the range of opinion and expert cross-examination of evidence, implementing a fully adversarial structure would mean recruiting specific panel members to advocate on behalf of a particular point of view, which may conflict with the goal of producing a consensus report. An example will be provided in section 2 that arises from U.S. National Academy of Science panel procedures. In addition, the rules themselves may simply not aspire to mimic adversarial argumentation, or may be ignored insofar as they do. In section 3, I will discuss how these concerns arise with respect to the operation of the Intergovernmental Panel on Climate Change (IPCC).

A related problem is that regulatory agency personnel are sometimes advocates for policy outcomes while at the same time working on the expert assessments that will guide decision-making. Examples are provided in compelling detail by Conrad and Ramey in their contributions to this volume.

Taken together these considerations point to a notable asymmetry regarding the safeguards imposed on court trials as opposed to expert panels. I therefore pose the question of why we would consider an adversarial system to be appropriate for a criminal trial, but not for a scientific assessment panel like the IPCC. My conclusion is that there would be merit in bringing more openly adversarial procedures into the IPCC.

This is based on two considerations. First, much of the trust people place in the IPCC is based on a belief that the process is more adversarial than is actually the case. For instance, there is a perception that the IPCC selects authors based on a competitive evaluation of credentials, when in fact the selection criteria are unknown. The only published criteria refer to geographical consideration (namely a balance between developed and developing countries). Also, there is a perception that the review process involves careful and objective weighing of conflicting arguments, but this is more of an ideal than reality, since the report-writing process puts authors in the position of assessing their own work (and that of their critics), and allows them to ignore or circumvent the comments of outside reviewers.

Second, the credibility of the final report is related to its claim of using peer-reviewed research, though in practice much non-peer-reviewed material is used. Notwithstanding that, there is wide variation in the meaning of the term "peer review." In the case of empirical research published in an academic journal, peer review does not typically involve anyone actually checking the data or calculations, nor does it even guarantee that such checking is possible. Hence the amount of scrutiny one expects evidence to undergo in an adversarial court trial cannot be assumed to have occurred when academic research is used in a policy-formation process, even when the research is "peer-reviewed." If policymakers want the evidence in a scientific assessment to have been as thoroughly examined as would be the evidence in a court trial, more adversarial procedures must be implemented.

II. THE U.S. NATIONAL ACADEMY OF SCIENCES EXPERT PANEL PROCESS

The U.S. National Academy of Sciences (NAS) operates under rules set out in the US Federal Advisory Committee Act.[1] In brief, they are aimed at ensuring that expert panels are not undermined by bias or conflicts of interest. When a panel is being created, the NAS nominates potential members, and the public is then invited to comment on the list. Potential members must report conflicts of interest, which can take the form either of financial interests or previous statements that would compromise the appearance of neutrality. By allowing for public comment on potential members the NAS creates a safeguard against some problems of bias or uniformity, since membership and composition can, to some extent, be contested (though not vetoed) by people outside the process.

The NAS requirements for balance recognize the existence of individual perspectives, but also try to rule out persons serving as advocates.

> The assessment of the necessary perspectives required for a particular study committee may also involve considerations that go beyond specific disciplinary scientific or technical concerns. For some studies, for example, it may be important to have an "industrial" perspective or an "environmental" perspective. This is not because such individuals are "representatives" of industrial or environmental interests, because *no one is appointed by the institution to a study committee to represent a particular point of view or special interest.* Rather it is because such individuals, through their particular knowledge and experience, are often vital to achieving an informed, comprehensive, and authoritative understand-

> ing and analysis of the specific problems and potential solutions to be considered by the committee. (www.nationalacademies.org/coi/bi-coi_form-0.pdf, page 3, italics in original)

The NAS recognizes that people may hold particular viewpoints, and in principle, it may be necessary to ensure that different points of view are represented in order to comprehensively examine a topic. Yet it is also assumed that, in his or her capacity as panel member, an expert is not a representative of a point of view. The two claims are somewhat contradictory. In effect the NAS is seeking the specialized knowledge an expert acquires as a result of having explored and defended a particular point of view over time, while at the same time asking the expert not to act as an advocate for that view. If we were to apply this principle to a courtroom we might have a situation in which the court calls for a generic "defense perspective" without actually assigning a lawyer to represent the specific defendant now on trial. The danger is this would end up either with a pretence of neutrality on the part of a lawyer who privately wants the defendant to prevail, or with the defense being conducted by someone who is indifferent to the defendant's cause but who is, nonetheless, believed by everyone else to be the person responsible for advocating it. In the latter case a situation might arise in which evidence and argumentation favorable to the defense fails to get placed before the court, and prosecution evidence receives inadequate cross-examination.

The NAS requires that panels hold public meetings, receive outside submissions, survey published literature, and conduct deliberations.[2] These provisions certainly increase the chance that the committee will consider all points of view. But the policies do not *require* it, and therein lies a big difference. At NAS hearings, witnesses have no right to cross-examine or question other witnesses. In a courtroom, because the defense attorney is present whenever evidence is presented, and has rights of cross-examination, it is very difficult for a court to avoid learning the defense attorney's view at every step along the way. It is, by contrast, quite possible for the NAS to overlook a point of view in the preparation of its report.

An example of this arose during the preparation of the 2006 NAS report entitled "Surface Temperature Reconstructions For the Last 2,000 Years" (North et al. 2006). The assessment had been undertaken at the request of the U.S. House of Representatives Science Committee in response to controversies over the so-called hockey stick graph (Mann et al. 1998, 1999; see Montford 2010 for a detailed chronology of the events leading up to the NAS panel being formed). The hockey stick had been prominently featured in the 2001 Report of the IPCC (IPCC 2001), and appeared to indicate that the Earth's climate steadily cooled for about nine hundred years, then began a sudden warming trend at around the start of the twentieth century. McIntyre and McKitrick (2003, 2005a,

2005b) had criticized the methodology and results, arguing that the underlying data could not support the authors' conclusions, nor those of the IPCC.

The hockey stick issue was a case where there were clearly opposing points of view to be considered. McIntyre and McKitrick were invited to present to the panel, as were many others, but as for the panel itself the only member who could be considered critical of the IPCC was John Christy of the University of Alabama-Huntsville. He was, himself, an IPCC Lead Author (LA), but he had also been critical of other aspects of its report.

In the final report (North et al. 2006) the NAS panel accepted all the technical criticisms of the hockey stick submitted by McIntyre and McKitrick, in particular agreeing that methodological errors biased the result and led to underestimation of the relevant uncertainties. Yet overall the committee deemed it "plausible" that the last few decades of the twentieth century were the warmest of the millennium in the Northern Hemisphere (North et al., p. 4), based on other studies published after 1998 that had yielded broadly similar conclusions as Mann's paper. It is at this point that the lack of cross-examination becomes relevant. The other studies cited by the NAS used much of the same data and, in some cases, the same methods as the Mann hockey stick, weakening their value as independent corroboration. But this point was not made by the NAS panel. McIntyre had anticipated this problem might arise and had pointed it out in his submission to the panel, but the chairman, Gerry North, later indicated that they had not considered the issue.[3] It is possible that, had a more adversarial form of cross-examination taken place, they would have done so, and the report's conclusions would have been different.

III. IPCC OPERATING PROCEDURES

A. Perceptions of Rigor

The high standing of the IPCC is due in part to the perception that the review process is detailed, adversarial and rigorous. In a 2007 interview, IPCC Chair Rajendra Pachauri described the process in these terms:[4]

> And then at each stage of the production of the report it goes through a process of reviews and all the comments we get are carefully logged, each one is taken into account. It is not necessary that everything will be accepted, but everything has to be considered . . . You know when we send out drafts or reviews, they go to everyone. They go to governments and governments can pass them out to all kinds of people. And if there are governments who have a cosy relationship with the naysayers, they would naturally send it to these people.

> And we take every single comment into account. We have a person called a review editor whose job it is to ensure that each particular comment is taken into account. So you know there is a kind of a monitor that takes care of the process being followed accurately.

When the United States Environmental Protection Agency (EPA) released its Proposed Endangerment Finding on greenhouse gases in April 2009 and its final Endangerment Finding in December 2010 it took the unusual step of not conducting any internal evaluation of the science, instead relying on, among other reports, the IPCC assessment, based on claims about its review process:

> [These] assessment reports undergo a rigorous and exacting standard of peer review by the expert community, as well as rigorous levels of U.S. government review and acceptance. Individual studies that appear in scientific journals, even if peer reviewed, do not go through as many review stages, nor are they reviewed and commented on by as many scientists. (U.S. Federal Register 74 page 66510–66511)

Similar hyperbole is easy to find in media coverage of the IPCC:

- The IPCC process is "arguably the most rigorous and transparent peer review process in the history of science."[5]
- "[The IPCC is] one of the most rigorous scientific review bodies in existence. Many thousands of scientists have dedicated their time to preparing and reviewing the most comprehensive and authoritative assessments of climate science available. In addition, governments from around the world have reviewed and approved the IPCC's key findings. The reports undergo two rounds of peer review, and the policymakers's summaries of the working groups are then subjected to a word-by-word approval of all governments in the presence of the chapter lead authors."[6]
- "Currently over 130 countries are taking part, with 450 authors and 800 contributors, while 2,500 scientific experts take part in the hearings."[7]
- "It is probably the single most exhaustively reviewed scientific document ever produced. It has been intensely scrutinised and few errors found."[8]

It is instructive to review the actual IPCC procedures in the context of such claims and perceptions. In the next two sections I will argue that the actual procedures are far less rigorous than such perceptions imply in some key respects. First, the author selection process is opaque and arbitrary. Second, the volume of review comments is uneven and is suggestive of widespread apathy among member countries. Third, the review process allows LAs the final say on the text, even over reviewer objections, and text can be inserted after the close of peer review.

B. Selection of Lead Authors and Review of Drafts

The IPCC Assessment Report process is managed by a thirty-member group called the IPCC Bureau, which consists of the IPCC Chair, several vice chairs, working group leaders and other key IPCC figures. Among other things, the Bureau selects the Lead Authors (LAs) of the assessment reports according to unpublished criteria. No advance announcement of the nominees is made; instead the IPCC simply publishes a list of its selections as a fait accompli. On its website, the IPCC describes the process for choosing LAs in rather unclear terms.[9]

> Coordinating Lead Authors and Lead Authors are selected by the relevant Working Group/Task Force Bureau, under general guidance and review provided by the Session of the Working Group or, in case of reports prepared by the Task Force on National Greenhouse Gas Inventories, the Panel, from those experts cited in the lists provided by governments and participating organisations, *and other experts as appropriate, known through their publications and works.* (emphasis added)

The highlighted phrase shows that the IPCC Bureau reserves the right to appoint Lead Authors from outside the list of those nominated by governments and participating organizations, based solely on the Bureau's view of their suitability, and no decision criteria are listed. The requirement for balance states only that the list of authors "shall reflect the need to aim for a range of views."[10]

The report of the InterAcademy Council (IAC 2010) on IPCC procedures focused on this lack of criteria as a source of concern to many observers.

> Although the individuals who corresponded with the Committee were generally supportive of the author teams chosen, few knew why some authors are chosen and others are not, and many found the selection criteria arbitrary. The absence of a transparent author-selection process or well-defined criteria for author selection can raise questions of bias and undermine the confidence of scientists and others in the credibility of the assessment . . . The IPCC has no formal process or criteria for selecting authors, although some Working Group Co-chairs established their own for the fourth assessment, considering factors such as scientific expertise and excellence, geography, gender, age, viewpoint, and the ability to work in teams. (IAC 2010, pp. 14–15)

The IAC recommended that "the IPCC should establish a formal set of criteria and processes for selecting Coordinating Lead Authors and Lead Authors." Pachauri's insistence, quoted above, that there are rigorous selection criteria is contradicted by the fact that the IAC finding that such criteria do not exist and need to be created.

Even if the author selection process were biased, it might be supposed that the review process keeps such biases in check. Drafts of IPCC reports are circulated to a long list of reviewers, including those openly critical of past IPCC conclusions. At no point in the process, however, are any of the reviewers asked whether or not they endorse the conclusions or the summary, which are written after the completion of the report-writing process. Also, there is a rewrite stage after the close of the peer review process. In the case of the Fourth Assessment Report of 2007, the final deadline for reviewer input on the second draft was in June 2006. Thereafter the report underwent another re-write in the fall of 2006, and further revision in the winter of 2007. These redrafts were not subject to expert review. The Summary for Policymakers was produced at a meeting of the IPCC in late January 2007, and that draft was not sent out for expert review or approval. Hence the involvement of people of diverse views during the review process does not imply that the final report or the summaries are constrained to reflect their views.

The IPCC website describes the review process further as follows.[11]

> In the course of the multi-stage review process, both expert reviewers and governments are invited to comment on the accuracy and completeness of the scientific/technical/socioeconomic content and the overall balance of the drafts. The circulation process among peer and government experts is very wide, with hundreds of scientists looking into the drafts to check the soundness of the scientific information contained in them. The Review Editors of the report (normally two per chapter) make sure that all comments are well taken into account.

This sounds impressive, but since all involvement with the IPCC is voluntary, no one is assigned the role of critically reviewing a particular section or chapter. It is conceivable that parts of a report might not be read by any reviewers at all: there is nothing in the IPCC procedures that prevents this.

The number of comments submitted is a good indication of how carefully each chapter was read. Although there are 195 governments in the IPCC, only twenty-two national governments submitted any review comments on the Second Order Draft (governments did not review the First Order Draft). The European Commission also contributed comments on two chapters, bringing the total to twenty-three government entities.[12]

Of the 2,010 comments submitted, over half were from only two countries: the United States and Australia. Not one African country submitted a comment, nor did any Middle Eastern or Arabic countries, nor did Russia nor the former Soviet states. Brazil submitted comments on three chapters and Chile commented on one chapter; other than that there were no comments from any South American countries. None of the small island states in the Pacific submitted comments. In eastern Europe, the Czech Republic commented on one chapter, and Hungary commented on three chapters; other than that there were no comments from any government in Eastern Europe. On the whole, the evidence is clear that, except for Australia and the U.S., government review was cursory or non-existent.

Chapter 3 of the IPCC Report contains the core material on observed climate change, and is the basis for many of the key conclusions. Only eleven countries—fewer than one in ten—submitted any comments. Chapter 9 is the most controversial, as it is the part of the report where the question of attribution to greenhouse gases is examined. Even fewer countries (nine) submitted comments, and of the 152 comments received, 113 were from the United States alone. Consequently, there is no evidence from the comment record that governments actually studied the core of the report in detail, with the exception of the United States and Australia, coincidentally the two non-signatories to the Kyoto Protocol at the time of the report's publication.

Now consider the expert review process itself. As shown by McLean (2009), one of the striking features of the IPCC drafts was the extent to which IPCC authors commented on their own chapters. For example, for chapter 9, there were fifty-six contributing authors and sixty-two reviewers. But of the reviewers, seven were also authors. So there were only fifty-five independent reviewers: in other words fewer reviewers than authors. This was the case for eight of the eleven chapters.

As mentioned, chapter 9 is critically important since it is the source of the claim that warming can be attributed to greenhouse gases. It is fair to say that without chapter 9 the remaining chapters would be of little policy interest, since they mainly document patterns of climate variability. Those patterns acquire political importance only if they can be attributed to human activity, and that question is addressed in chapter 9. The scrutiny given to that chapter is a key measure of the rigor of the overall IPCC process and the soundness of its main conclusions.

Of the approximately 3,500 experts the IPCC claims contributed to the Fourth Assessment Report, only sixty-two contributed review comments on the Second Order Draft of chapter 9 of Working Group I. Of these, as already noted, seven were also authors. Also, three were editors of the IPCC Report, one was an employee of the IPCC Technical Support Unit, and twenty-six were authors or coauthors of papers discussed in the chapter. Ten of the reviewers advocated on behalf of their own papers in their review comments. Only thirty-one reviewers could be identified as truly independent (McLean 2009). While the views of these reviewers may not have been fully expressed, only four explicitly stated support for the overall chapter. Assuming more supported the chapter but did not say so, it is still the case that only a few dozen independent reviewers read the chapter and agreed with its conclusions. Moreover, of the sixty-two reviewers, more than half contributed only one or two comments, suggesting they did not even read the whole thing.

This does not necessarily mean that a lot of scientists disagree with chapter 9 or with the IPCC Report as a whole. But it does show that the core of the report received relatively little review.

As to the question of whether LAs are required to deal with adversarial review comments, the IPCC points to the existence of Review Editors (REs) as a safeguard, but rarely acknowledges that LAs, not REs, have the final say on the text. They must make a response to comments, but their responses are not part of the report itself, and up until recently they were not even publicly released. REs are only supposed to examine them to ensure that due consideration has been given to comments.[13]

Many comments are simply rejected. For instance, over half the comments on chapter 9 of the Second Order Draft were rejected. In cases in which an LA wants simply to reject a comment, the RE can advise on the matter, but cannot over-rule the LA.

The IPCC rules say[14] that if there are significant unresolved differences of opinion, REs must supply annexes that explain them. That rule, however, is apparently a dead letter, since no such annexes have ever been produced. One of the most contentious disputes between LAs and reviewers concerned the paleoclimate chapter, in particular the hockey stick. As an IPCC reviewer, and a coauthor of the papers at the center of the dispute, I can attest to the deep disagreement between reviewers and LAs on how this matter should be presented, as is shown in the review record. Yet no annex was produced, and the REs apparently signed off on the chapter nonetheless. In 2008, a U.K. citizen named David Holland sought information about how one of the chapter 6 REs (John Mitchell of the U.K. Met Office) had handled the controversies over the Mann et al. hockey stick. As part of his inquiries, Holland submitted a Freedom of Information Act request to the Met Office. The documents released in reply contained an email from IPCC Chair Susan Solomon to Mitchell advising him on the limitations of his responsibilities to provide explanations for editorial decisions. The email was dated March 14, 2008, and stated, in part, the following:

> The review editors do not determine the content of the chapters. The authors are responsible for the content of their chapters and responding to comments, not REs. Further explanations, elaboration, or re-interpretations of the comments or the author responses, would not be appropriate.[15]

Taking these points together, it is clear that REs do not have the authority to stop LAs who are determined to make arbitrary decisions about chapter content. For this reason the IPCC review process is fundamentally unlike the academic peer review process, in which the editor reserves the final right to accept or reject a paper and its contents based on review comments.

C. Changes to the Text Outside the Review Process

The IPCC rules do not prevent inserting or deleting material after the close of review, or using unpublished material. An example of an unwarranted deletion arose in chapter 3 around the issue of long term persistence.[16]

In this case, text was introduced into the Second Order Draft of the report, based on expert reviewer comments on the First Draft and supported by citations to peer-reviewed literature, that expressed caution about the statistical significance of warming trends in climate data. Despite the fact that the text had been agreed to during the review phase, it was then deleted after the close of scientific review, prior to final publication.

The underlying issues are technical. While it is relatively straightforward to estimate a linear trend through time series data, it is much more difficult to determine if it is statistically significant when the data exhibits a strong form of autocorrelation called "persistence." The statistical literature discusses a related family of concepts called, variously, long memory, long term persistence (LTP), persistency/antipersistency (P/AP), autoregressive integrated moving averages (ARIMA), fractional integration (ARFIMA), and so forth. A large literature exists on each of these topics and many papers have been published applying the estimation methods to climatology data sets. Indeed many of the foundational works come out of the hydrology field, where LTP models were developed by pioneers like Hurst and Mandelbrot to provide more physically realistic characterizations of long data sets (see Koutsoyiannis 2002).

One of the established results of the LTP literature is that taking it into account tends to reduce the apparent significance of trends in long data sets. This has been shown in temperature analysis as well as in analysis of other data sets (e.g. Cohn and Lins 2005). The First Draft of the IPCC report chapter 3 contained no discussion of this topic and also made some strong claims about trend significance based on unpublished calculations done at the CRU. I was one of the reviewers who called attention to the issue and requested insertion of some cautionary text dealing with the LTP issue. Chapter 3 was revised so that the Second Draft now included a caution (on pages 3–9) that "many oversimplified techniques will tend to overstate the significance" of trends, and failure to deal with persistence "can make erroneous detection of trends a typical outcome in climatic data analysis."

Then, after the close of Expert Review, this text was deleted and replaced with wording that implied the IPCC methods were suitable for computing trends. There had been no comments from reviewers requesting such changes. The science that led to the original inclusion of the cautions was in good quality peer-reviewed journals, the chapter authors had agreed to its inclusion during the review process and there were no reviewer objections to its inclusion. But, evidently, at some point, one or more of the LAs decided they did not want to include it anymore, and the IPCC rules do not prevent arbitrary deletion of material even after it has been inserted as a result of the peer review process.

An example of a post-review insertion concerns evidence of contamination of the surface temperature record. The IPCC released the First Order Draft in August 2005. The relevant section was chapter 3, pages 3–10, in which there was no mention of studies by McKitrick and Michaels (2004) or de Laat and Maurellis (2004) that had concluded the surface data used by the IPCC had warming bias due to industrialization in the regions where temperatures were monitored. IPCC Expert Reviewer Vincent Gray criticized the omission, as did I.[17]

The IPCC Second Order Draft was released in March 2006. Again there was no mention of the findings, even though a new study had appeared (de Laat and Maurellis 2006) confirming the earlier ones. Reviewers provided extensive criticism of this oversight.

In June 2006 the expert review period closed. The final, published IPCC report in May 2007 included a new paragraph in chapter 3, on page 244, that had not been included in either of the drafts shown to reviewers. It read (emphasis added):

> McKitrick and Michaels (2004) and De Laat and Maurellis (2006) attempted to demonstrate that geographical patterns of warming trends over land are strongly correlated with geographical patterns of industrial and socioeconomic development, implying that urbanisation and related land surface changes have caused much of the observed warming. However, the locations of greatest socioeconomic development are also those that have been most warmed by atmospheric circulation changes (sections 3.2.2.7 and 3.6.4), which exhibit large-scale coherence. *Hence, the correlation of warming with industrial and socioeconomic development ceases to be statistically significant.* In addition, observed warming has been, and transient greenhouse-induced warming is expected to be, greater over land than over the oceans (chapter 10), owing to the smaller thermal capacity of the land.

The concept of "statistical insignificance" has a specific numerical interpretation: it implies that an empirical test has been done of a null hypothesis yielding a p value greater than 0.1. The effects reported in McKitrick and Michaels (2004) had p values on the order of 0.002 or 0.2 percent, indicating statistical significance of the effects. The claim in the final IPCC Report that the correlation between warming and socioeconomic and industrial development was statistically insignificant was made without any supporting evidence. There is no study that demonstrates such a lack of statistical significance; quite to the contrary, McKitrick (2010) was later published disproving the specific claim that accounting for atmospheric circulation effects renders the earlier results insignificant.

This is not a peripheral topic: the papers in question point to contamination of the surface temperature data on which most of the main IPCC conclusions depend, including the attribution of climate change to greenhouse gases. The claim that work showing such contamination lacks statistical significance—the highlighted portion of the above paragraph—was not shown to expert reviewers during the IPCC Report preparation. Moreover, the references to sections 3.2.2.7 and 3.6.4 of the IPCC Report are misleading since neither section presents evidence that warming due to atmospheric circulation changes occurs in the regions of greatest socioeconomic development. Neither section mentions industrialization, socioeconomic development, urbanization, or any related term. The final sentence in the quoted paragraph is irrelevant to the present discussion since the debate only concerns data over land: there is obviously no economic development over the open ocean.

This illustrative sequence of events shows that nothing in the IPCC procedures prevents arbitrary removal and insertion of material after the close of the scientific review phase, and nothing prevents a chapter author from making unsubstantiated claims in order to discredit and discount published scientific work that is inconsistent with a Lead Author's preferred conclusions.

D. Summary

The IPCC operates according to procedures that differ in subtle but important ways from the common perception. There are thousands of people involved in writing and reviewing the IPCC reports, but nobody is actually assigned the task of reviewing any section in particular, and many sections, including key chapters on which the politically prominent conclusions are based, may only receive cursory review. Most governments made no comments on the drafts and hence there is no evidence that they read the contents at all. While the scientific review phase involved two rounds of expert review, LAs thereafter had a free hand to make arbitrary changes to their chapters outside the purview of the reviewers, even if the changes contradicted the results of earlier review. Review Editors had no authority to overrule LAs, even when review comments were rejected, or material previously inserted on reviewer instruction was subsequently removed, or new and unsubstantiated material was inserted outside the peer review process.

It is hardly necessary to spell out the contrast between these procedures and those used in courts. The IPCC process is like one in which the prosecutor is also the judge. Also, in the IPCC analogy, while the defense is allowed to speak, the judge is allowed to leave the room during the defense presentation, and the prosecution is allowed to present new evidence to court after the defense has gone home for the day. If the IPCC were reformed to adopt adversarial, courtroom-like processes, it would imply that deliberate steps would have to be taken to allow proponents of alternate points of view to debate the evidence, LAs would not be able to rewrite the text after the close of the review process, and the conclusions of the report would only be arrived at after material had received independent, adversarial scrutiny.

IV. PEER REVIEW REQUIREMENTS

Another way in which the assumption of adversarial review is applied to expert panel processes is the concept of "peer review." As noted in section 3.1 above, the term is used repeatedly as evidence that deep and rigorous review of evidence occurs during the IPCC process. Controversies arose in the winter of 2010 when it was discovered that the IPCC had made use of non-peer reviewed material as the basis for a number of its claims, such as the rate of melting of Himalayan glaciers.[18]

However, had they confined themselves to so-called peer-reviewed literature it would not be a guarantee of quality. The process by which papers go into scientific journals does not necessarily involve rigorous checking or scrutiny. Journal referees almost never check data or calculations, and journals rarely require authors to archive their data and code, making it difficult for others to check their work. The appearance of a paper in a journal only means that referees looked at it and recommended to the editor that it be published, and the editor accepted the recommendation. It does not mean that the publisher, editor, or reviewers provided any guarantee of accuracy.

McKitrick and McCullough (2009) examined cases across multiple disciplines in which policy-relevant research was published, which later turned out to be unreproducible because the authors would not share their data or disclose their methods. McCullough and his colleagues have also conducted replication audits at numerous economics journals, most prominently the *Journal of Money, Credit and Banking* (*JMCB*), and the *American Economic Review* (*AER*). A 1986 audit of the *JMCB* had earlier showed that only two of fifty-four sampled articles could be replicated, even when the authors were cooperative (and many refused to cooperate). In response, the journal adopted a policy requiring archiving of data and code, but thereafter it was rarely enforced. A follow-up audit by McCullough et al. (2006) showed that fewer than one-third of the empirical papers had archive entries, and of the sixty-two selected for examination only fourteen could be replicated. Similar problems were exposed at the *AER* and other prominent journals.

McKitrick and McCullough (2009) documented cases from other disciplines, including ecology, medicine, public health, forestry, finance, and history, in which independent replication of important studies was actively thwarted through the refusal to share data. The problem is pervasive and widely recognized among academics. Possibly as a result of the ubiquitous nature of the problem, independent replication of published research is extremely rare. To some extent this is not a serious problem, since most published academic papers are unimportant. Academics generally understand that publication is not a guarantor of quality, it is in some sense just the beginning of the review process, and published results should only be considered provisional until they have been replicated and reproduced by others.

The problem arises when users of academic research believe that the term "peer review" means someone has checked the results and has independently verified the conclusions. Looking at the IPCC, a further problem arises when LAs are in the position of summarizing not only their own research, but also work they (or a close colleague) reviewed for journals. If an IPCC LA refereed an article for a journal, the fact that it was published in a "peer-reviewed" journal does not provide independent attestation as to is fitness for use by the IPCC.

To give an example, consider the following email dated May 20, 2005, from Phil Jones to Michael Mann. Jones was the head of the Climatic Research Unit (CRU) at the University of East Anglia and had been appointed as LA for the next IPCC report (which would appear as IPCC 2007). Mann is a climatologist at Penn State and had been LA for the previous IPCC report. Jones's CRU colleague Keith Briffa was also an LA for the 2007 IPCC report, with responsibility for the paleoclimate chapter. At the time, papers were being published that questioned Mann's "hockey stick" graph (see section 2 above). In his capacity as IPCC LA, Mann had highlighted the graph in the 2001 IPCC report. But the most recent papers in the journals had concluded that the graph was flawed, and Briffa would need to summarize these findings, which might prove embarrassing for the IPCC. Phil Jones had been invited by the journal *Climatic Change* to review a submission by Mann's former student Caspar Ammann and coauthor Eugene Wahl, which argued that the McIntyre and McKitrick criticisms of the hockey stick were invalid. Jones wrote to Mann as follows ("MM" refers to McIntyre and McKitrick).[19]

> Mike, Just reviewed Caspar's paper with Wahl for Climatic Change. Looks pretty good. Almost reproduced your series and shows where MM have gone wrong. Should keep them quiet for a while. Also they release all the data and the R software. Presume you know all about this. Should make Keith's life in Ch 6 easy!

The paper Jones reviewed so enthusiastically was subsequently rejected. Ammann and Wahl's submission to *Climatic Change* did not include the key calculations; instead, it relied on calculations in another paper they had submitted to *Geophysical Research Letters*. But that paper had also been rejected, indeed prior to the submission to *Climatic Change*. This was pointed out by the other reviewer (Stephen McIntyre), who also asked to see some test scores that Ammann and Wahl had left out of their submission. Ammann and Wahl refused to produce the test scores (see Montford 2010, chapter 8 for the detailed history). So the *Climatic Change* submission was rejected. A much different version was later submitted and accepted, but not until 2007, and it contained some important caveats regarding the hockey stick verification scores.

These two reviewers—Jones and McIntyre—could rightly be characterized as holding diametrically opposed views in the matter of the Ammann and Wahl submission. Jones liked the conclusions and was favorably disposed towards the paper. He was a friend of Michael Mann, whose work Ammann and Wahl were trying to defend, as well as a close colleague of Keith Briffa, whom he perceived would find his IPCC assignment easier to fulfill if there was something in print that rebutted McIntyre's critique of the hockey stick. So Jones, being especially alert for material that appeared to support his biases, endorsed the Ammann and Wahl paper for publication.

McIntyre, by contrast, doubted Ammann and Wahl's conclusions, and requested the underlying data and calculations. It turned out that Ammann and Wahl had reproduced the same numerical results that McIntyre and McKitrick had in one of their papers (McIntyre and McKitrick 2005b), but they proposed a different interpretation. McIntyre knew the flaws in that interpretation and knew which calculations to examine to show why the interpretation was untenable. When he looked for the calculations, he found they were either cited as being in the *Geophysical Research Letters* submission, which he knew had already been rejected, or were not discussed at all. Once McIntyre brought these matters up the paper could not proceed, even though it would be accurate to say that the editor of *Climatic Change*, Stephen Schneider, shared Jones's overall outlook on climate issues.

The adversarial position of Jones and McIntyre in the refereeing process might be taken to mean that the review process "worked." But it is the exception that proves the rule, since in this one case opposing "counsel" were on equal footing. Nothing guarantees that journal review processes will always operate that way. It was the Editor's decision to ask McIntyre to provide a referee report, but nothing guarantees that an Editor will solicit opposing views. In this one instance, Jones and McIntyre were on equal institutional footing, and McIntyre won the case. But thereafter Jones held the role of IPCC Lead Author with a free hand to write his chapter, and McIntyre, as a mere reviewer, had no standing to challenge or determine the text. And despite publishing numerous subsequent papers on paleoclimate reconstructions, neither *Geophysical Research Letters* nor *Climatic Change* ever asked McIntyre to serve as a reviewer again.

V. INTRODUCING ADVERSARIAL ELEMENTS INTO EXPERT PANELS

The adversarial system in court is by no means perfect, but it does have the advantage that evidence will be presented for each side and claims will be subject to cross-examination. If a courtroom operated like an expert panel, trials would look very different. Taking the IPCC as an example, the court would appoint the people it wants to write the judgment, and if a pattern emerged in which advocates for a particular point of view kept getting appointed to write the judgments, there would be no recourse. While there would be a nominal process of examination of evidence, there would be no guarantee of adversarial scrutiny nor right of cross-examination. To the extent scrutiny of evidence and argumentation did occur, it would be followed by one or more stages in which evidence was presented to judges in private without defense counsel present, and judges would have the right to strike defense evidence from the record after the presentation of final arguments.

What would be different if an expert panel like the IPCC had to function more like a court? The first and most significant change would be that specific members would be sought whose responsibility would be to put opposing viewpoints forward, and specific sections of the report would be given over to allowing them to present their evidence. Suppose the text of the Assessment Report contains a section arguing for a specific idea, such as that the sun has relatively little influence on the climate. Then it would also contain a section presenting evidence in the other direction. And the respective authors would have a chance to respond to each other's sections. Indeed the whole report could be structured this way, as a series of arguments, counterarguments, and responses.

Who, then, would "decide" the verdict? In the case of the IPCC, it ought not to matter. For a long time we have been told that the evidence is overwhelmingly strong in one direction and the matter is settled, so presenting both sides should only serve to emphasize this. But if presentation of both sides were actually to lead a reader to conclude that it is not obvious which one is right, such a reader would likely not want a third party to issue a summary verdict anyway.

At present, the IPCC Summary is released months before the actual report, and the authors of the Summary are able to impose their interpretation because nobody is in a position to check what they say against the report, much less the underlying science, until long after media and political attention has faded. But suppose an adversarial panel report came out without a verdict, and a government agency were then asked to produce a summary and an overall conclusion: they would be working within the constraints imposed by the fact that the arguments and data would already be in circulation, and if the summary failed to do justice to the substance it would be promptly discredited.

In practice, an adversarial process makes it impossible for the report to be modified or rewritten after external review is over. The text would have to be produced iteratively with LAs and reviewers exchanging text under the supervision of a Review Editor, until they either reach agreement or agree to a separate annex describing the unresolved disagreement, at which point the text would be frozen and no further changes would be permitted.[20]

In this way, the IPCC report might end up containing a number of sections in which conflicting evidence and arguments are presented, with no attempt to resolve the disputes.

This would create an interesting pair of possibilities. One possibility would be that one case clearly dominates the other, and any open-minded reader would look at the report and see which side has the better evidence. The other possibility would be that each side has a good case and it is not clear which one is correct. Would that be a problem? No—if that is where matters stand, that is what the report should say. A reliable process applied to inconclusive evidence should report that the evidence is inconclusive. It is not an indicator of the reliability of a process that it always delivers a unanimous verdict irrespective of the evidence. Were that always to happen we would have reason to believe the process is not entirely trustworthy.

NOTES

1. See www.nationalacademies.org/coi/bi-coi_form-0.pdf.
2. See www.nationalacademies.org/studycommitteeprocess.pdf.
3. See climateaudit.org/2009/04/19/epa-quality-guidelines-and-the-nas-panel/.
4. See www.rediff.com/news/2007/jun/05inter.htm.
5. Paul Beggs, Macquarie University, quoted in insideclimatenews.org/news/20100226/climate-scientists-defend-ipcc-peer-review-most-rigorous-history.
6. Robert Watson, former IPCC Chair, quoted in www.webcitation.org/5wcFvcZJU.
7. Professor Ole Danbolt Mjøs, Chairman of the Nobel Peace Prize Committee, on awarding the IPCC the Nobel, quoted in www.webcitation.org/5wakBqzkM.
8. Penny Wong, Australian Federal Minister for Climate Change and Water, cited in www.webcitation.org/5wcp0Rmjh.
9. See ipcc.ch/pdf/ipcc-principles/ipcc-principles-appendix-a.pdf Section 4.2.2.
10. See ipcc.ch/pdf/ipcc-principles/ipcc-principles-appendix-a.pdf Section 4.2.2.
11. See ipcc.ch/organization/organization_procedures.htm.
12. See McLean (2009).
13. See ipcc.ch/pdf/ipcc-principles/ipcc-principles-appendix-a.pdf section 5.
14. See ipcc.ch/pdf/ipcc-principles/ipcc-principles-appendix-a.pdf section 5.
15. David Holland, pers. comm.
16. The next few paragraphs draw upon my submission of evidence to the Muir Russell Inquiry rossmckitrick.com/uploads/4/8/0/8/4808045/mckitrick-iccer-evidence.pdf.
17. pds.lib.harvard.edu/pds/view/7795947?n=7&imagesize=1200&jp2Res=.25.
18. See, for example, ipcc.ch/pdf/presentations/himalaya-statement-20january2010.pdf.

19. yourvoicematters.org/cru/mail/1116611126.txt.
20. For more explanation of this option see McKitrick (2011b).

REFERENCES

Briffa, K. R., 2000. "Annual Variability in the Holocene: Interpreting the Message of Ancient Trees." *Quaternary* 43 *Science Reviews*, 19:87–105.

Brohan, P., Kennedy, J., Harris, I., Tett, S. F. B., and Jones, P. D. (2006). "Uncertainty Estimates in Regional and Global Observed Temperature Changes: A New Dataset from 1850." *J. Geophys. Res.* 111, D12106, doi:10.1029/2005JD006548.

Cohn, T. A., and H. F. Lins (2005). "Nature's Style: Naturally Trendy." *Geophysical Research Letters* 32, DOI: 10.1029/2005GL024476.

De Laat, A. T. J., and A. N. Maurellis (2004). "Industrial CO_2 Emissions as a Proxy for Anthropogenic Influence on Lower Tropospheric Temperature Trends." *Geophysical Research Letters* vol. 31, L05204, doi:10.1029/2003GL019024.

De Laat, A.T.J., and A.N. Maurellis (2006). "Evidence for Influence of Anthropogenic Surface Processes on Lower Tropospheric and Surface Temperature Trends." *International Journal of Climatology* 26: 897–913.

InterAcademy Council (IAC) (2010). "Climate Change Assessments: Review of the Processes and Procedures of the IPCC." Amsterdam: The IAC.

IPCC (2001). *Climate Change 2001: The Physical Science Basis. Contribution of Working Group I to the Third Assessment Report of the Intergovernmental Panel on Climate Change.* Cambridge: Cambridge University Press.

IPCC (2007). *Climate Change 2007: The Physical Science Basis. Contribution of Working Group I to the Fourth Assessment Report of the Intergovernmental Panel on Climate Change*, edited by S. Solomon, D. Qin, M. Manning, Z. Chen, M. Marquis, K. B. Averyt, M. Tignor, and H. L. Miller. Cambridge: Cambridge University Press.

Jones, P. D., and A. Moberg (2003). "Hemispheric and Large-Scale Surface Air Temperature Variations: An Extensive Revision and an Update to 2001." *Journal of Climate* 16: 203–23.

Jones, P. D., M. New, D. E. Parker, S. Martin, and I. G. Rigor (1999). "Surface Air Temperature and Its Changes Over the Past 150 Years." *Reviews of Geophysics*, 37: 173–99.

Jones, P. D., Briffa, K. R., Barnett, T. P., and Tett, S. F. B. (1998). "High-Resolution Palaeoclimatic Records for the Last Millennium; Interpretation, Integration, and Comparison with General Circulation Model Control-Run Temperatures," *The Holocene* 8: 455–71.

Koutsoyiannis, D. (2002). The Hurst Phenomenon and Fractional Gaussian Noise Made Easy. *Hydrological Sciences Journal* 47, 4: 573–95.

Mann, M. E., Bradley, R. S., and Hughes, M. K. (1998). "Global-Scale Temperature Patterns and Climate Forcing Over the Past Six Centuries." *Nature* 392: 779–87.

Mann, M. E., Bradley, R. S., and Hughes, M. K. (1999). "Northern Hemisphere Temperatures During the Past Millennium: Inferences, Uncertainties, and Limitations." *Geophysical Research Letters* 26: 759–62.

McCullough, B. D., Kerry Anne McGeary, and Teresa D. Harrison (2006). "Lessons from the JMCB Archive." *Journal of Money Credit and Banking* 38(4): 1093–1107.

McIntyre, Steven, and Ross McKitrick (2003). "Corrections to the Mann et al." (1998) "Proxy Data Base and Northern Hemisphere Average Temperature Series." *Environment and Energy* 14(6): 751–71.

McIntyre, Stephen, and Ross McKitrick (2005a). "Hockey Sticks, Principal Components and Spurious Significance," *Geophysical Research Letters* vol. 32, no. 3, L03710 10.1029/2004GL021750 12 February 2005.

McIntyre, Stephen, and Ross R. McKitrick (2005b). "The M & M Critique of the MBH98 Northern Hemisphere Climate Index: Update and Implications." *Energy and Environment* 16(1): 69–100.

McKitrick, Ross R. (2011a). "Bias in the Peer Review Process: A Cautionary and Personal Account" in Michaels, Patrick J. ed., *Climate Coup*. Washington, DC: Cato.
McKitrick, Ross R. (2011b). "What is Wrong with the IPCC? A Proposal for Radical Reform." London: The Global Warming Policy Foundation, November 2011.
McKitrick, Ross R. (2010). "Atmospheric Oscillations do not Explain the Temperature-Industrialization Correlation." *Statistics, Politics and Policy,* vol 1 no. 1, July 2010.
McKitrick, Ross R., and P. J. Michaels (2007). "Quantifying the Influence of Anthropogenic Surface Processes and Inhomogeneities on Gridded Global Climate Data." *Journal of Geophysical Research* 112, D24S09, doi:10.1029/2007JD008465.
McKitrick, Ross R., and Patrick J. Michaels (2004). "Are Temperature Trends Affected by Economic Activity? Reply to Benestad (2004b)" *Climate Research* 27(2):175–76.
McKitrick, Ross R., and B. D. McCullough (2009). "Check the Numbers: the Case for Due Diligence in Policy Formation." Fraser Institute, February 2009.
McLean (2009). "Peer Review? What Peer Review? Failures of Scrutiny in the UN's Fourth Assessment Report." Washington, DC: Science and Public Policy Institute.
Montford, Andrew (2010). *The Hockey Stick Illusion*. London: Stacey International.
North, G., et al. (National Research Council, NRC) (2006). *Surface Temperature Reconstructions for the Last 2,000 Years*. Washington, DC: National Academies Press.

II

Taxonomy and Endangered Species Regulation

FOUR

On The Origin of Specious Species

Rob Roy Ramey II

> Species: A term which everybody thinks they understand, but which nobody agrees upon, to denote the 'basic units' of groups of biological organisms.
> —Dr. John S. Wilkins, *Evolving Thoughts* science blog (2007)

The definition of "species" has been a subject of debate for over 2,300 years, since the time of Aristotle, and there is no end in sight (Hull 1965; Darwin 1859; Hendry et al. 2000; Marris 2007). That fact does not bode well for its use in present-day regulatory science.

Taxonomy was once the obscure domain of museum curators and naturalists, but times have changed. Expanding human populations have led to extinction or endangerment of many organisms worldwide. Legitimate concerns over this loss of biodiversity have led to the passage of laws and treaties that regulate the impact of human activities on species, subspecies, and populations bringing new legal, financial, and social relevance to the classification of all organisms. Although the first species-based protection laws primarily regulated hunting and harvests, later laws focusing on endangered species protection have had much broader implications and cover a greater diversity of taxa. These include the U.S. Endangered Species Act (ESA) of 1973, the Convention on International Trade in Endangered Species of Wild Fauna and Flora (CITES) of 1973, Australia's Environment Protection and Biodiversity Conservation Act of 1999, Canada's Species at Risk Act of 2002, and South Africa's National Environmental Management: Biodiversity Act of 2004. The scope of protection from each of these laws and treaties also extends to subspecies and a sub-subspecies category of "distinct populations." There are two types of "distinct populations" that can be listed under the ESA: the

Distinct Vertebrate Population Segment (DPS) and the Evolutionarily Significant Unit (ESU). DPS is a category applied to terrestrial vertebrates and ESU's are a special category used for Pacific salmon. In ESA recovery plans, the U.S. Fish and Wildlife Service (USFWS) also make use of two further subdivisions of species and subspecies in its implementation: management unit (MU) and recovery unit (RU). These are population to subpopulation-level categories that may have different management prescriptions and recovery criteria.

The descriptions of many of the species and subspecies listed as threatened or endangered predate these laws and were based on information that would typically not be acceptable today: small sample sizes, unknown inheritance of traits, and little or no quantitative basis or statistical analysis. Many are largely based on opinion. Essentially, a species or subspecies was what a taxonomist said it was. While modern taxonomists typically rely on much more sophisticated tools, including DNA sequence analysis and larger sample sizes, a substantial body of recent work suffers from the same epistemological shortcomings as in the past: a reliance on post hoc interpretations of data and gross discrepancies in how taxonomic assignments are made (Avise and Johns 1999; Zink 2004). The requirement for physical evidence (specimens) to support a species description has also been variously interpreted: in a recent case, a "new" species was described from photographs rather than the physical specimens (Timm et al. 2005).

The ESA's reliance on species and subspecies taxonomies means that the subjective opinions of taxonomists and their supporters can carry the force of law, influence the expenditures of hundreds of millions (and potentially billions) of dollars for the protection and recovery of taxa listed as threatened or endangered, and affect the future direction of conservation efforts worldwide. The near-omnipotence of the ESA has catapulted taxonomy from an obscure science to one of relevance for nearly all aspects of society in the United States: everything from housing prices to water supply.

Although improved methods have advanced our understanding of evolutionary processes in the natural world, these advances in genetics, morphometrics, behavior, bioacoustics, physiology, and ecology have also led to evolution in the definitions of species and contributed to taxonomic revision. Species concepts have also proliferated (twenty-seven by one count), yet all have been hindered in some way by vague definitions, notable exceptions, or issues of practicality. Although most of these are derivatives of the biological species concept (Mayr 1970) or the phylogenetic species concept (see Agapow et al. 2004 for a review), none has proven to be universally applicable and some authors have advocated abandoning the term *species* altogether (Hendry et al. 2000). Like Linnaeus (1735), the founder of modern taxonomy, taxonomists continue to contend with the dilemma of trying to draw lines through continua of natural variation and are often faced with conflicting information from different data sets. Thus, species and subspecies can be frequently be "in the eye of the beholder."

TAXONOMIC INFLATION

Following Linnaeus's work in the late 1700s, as more species were described, there was an increasing tendency to split existing species into several additional "new" species. Describing new species brought prestige to taxonomists, honor to those who they were named after, and expansion to specimen collections. This trend may have reached its high point with Merriam's 1918 description of eighty-two species of North American brown bears. Subsequently, the category of "subspecies" was introduced as a way to describe geographic variation often found within species, and as a result, many of those newly described species were demoted to the rank of subspecies. The addition of this subspecific level of classification resulted in the trinomial genus-species-subspecies nomenclature still in use. (A useful rule of thumb for distinguishing between species and subspecies: if there are two Latin names, it is a species; if there are three, it is a subspecies. Plant taxonomists also use the term: "varieties" (varietas) as a sub-subspecies category, so the last Latin name in the trinomial will be proceeded by sspp. or var. to distinguish these categories.)

In the early to mid-1900s, subspecies designations became a fertile ground for "splitting" into additional subspecies. Many new subspecies were described, and frequently these designations were based on minor superficial traits having no known genetic basis or were based on small sample sizes. Subspecies were split further into additional subspecies. During this time, the number of recognized species actually *declined* as some were reclassified as subspecies. The high tide of the subspecies trend was exemplified by Hall and Kelson's (1959) recognition of 214 subspecies of southern pocket gopher (*Thomomys umbrinus*) in the southwestern United States and northern Mexico. Practically every valley held a different subspecies. These same authors reported that the number of North American mammal species decreased from 1,399 in the year 1923 to 1,003 in the year 1957, whereas the number of subspecies more than doubled, from 1,155 to 2,676, during the same period.

In more recent decades, this trend has reversed, and an increasing number of "new" species are being described from former subspecies because of changing definitions. As noted by Meiri and Mace (2007), "the total number of mammal species [worldwide] has risen from 4,659 in 1993 to 5,418 in 2005." Yet, this trend is not the result of a renaissance of exploration and discovery in biological hotspots of unexplored wilderness; it is instead primarily attributable to many subspecies and populations being elevated to the level of species, a process known as "taxonomic inflation."

Taxonomic inflation has the effect of both increasing the perception of widespread endangerment while devaluing the basic currency of conservation: species. That is because the more that species are subdivided into subspecies and sub-subspecies categories, the smaller their ranges will be and the smaller the number of individuals found in the supposed endangered "population." Thus, the more endangered each appears to be (see Agapow et al. 2004). The more "species" that are perceived to be in trouble, however, the less money is available for helping each one of them. The United States allocates the majority of its endangered species budget to nondistinct but presumably threatened or endangered populations of common species, listed as DPSs and ESUs (Ramey et al. 2005; Marris 2007). It is clear, however, that this conservation approach comes at the expense of many "full" species (highly unique and readily distinguishable from other species) that are far more endangered (e.g. California condor, blue whale). With many full species endangered worldwide, and limited resources to save them, other nations may not find the ESA model to be a desirable or sustainable approach to conservation. Fully one-fifth of ESA-listed "species" are really subspecies or DPSs (Ramey et al. 2005), and these tend to be the most controversial and costly to protect (e.g., an abundance of Pacific salmon ESUs, gray wolf DPSs, and presumed subspecies of otherwise widespread species of meadow jumping mice, beach mice in Florida, and California gnatcatchers).

In 2007, *The Economist* published an editorial opinion and *Nature* published a lengthy news feature, both commenting on the implications of an upward trend in taxonomic inflation following the three-hundredth birthday of Linneaus (Marris 2007; The Economist 2007). These events raise the question: Why has taxonomic inflation reached a point that scientists and economists have begun to take notice? First, the application of the phylogenetic species concept (PSC) has led to the splitting of taxa into smaller "diagnosable" units, if each shares at least one unique heritable trait (a shared derived character providing evidence of monophyly). This method infers an evolutionary tree from genetic data and considers species to be those groups of organisms making up the terminal branches of the tree. Thus, even a single nucleotide substitution in DNA that is shared among all individuals on a terminal branch may be sufficient evidence of species status, even in the absence of any other distinguishing characteristic. Widespread use of genetic data in conservation genetics, especially relatively easy to obtain mitochondrial DNA sequence data, has led to broader application of the PSC (Agapow et al. 2004) and consequent "splitting." Subspecies are not typically a part of this scheme because PSC purists consider any diagnosable unit a full species. However, some PSC practitioners, particularly those involved in promoting listings under the ESA, have applied a less rigorous standard, by describing species or subspecies that are not strictly monophyletic or are sampled in such a way that the otherwise shared variation in intervening populations is excluded from consideration. This tactic is known as "isolation by sampling design." The endangered listing of the Preble's meadow jumping mouse was rationalized in this way by King et al. (2006) despite contrary data from morphology, ecology, and genetics (see Ramey et al. 2007 for a response).

A second contributing factor to taxonomic inflation can be an assumption of reproductive isolation under the biological species concept (BSC) using weak or unsubstantiated data. Admittedly, testing for reproductive isolation between presumed species, especially in the wild, has been a challenge for the practical application of the BSC (e.g., when the presumed species exist on different continents, making breeding studies impractical). Such constraints necessitate an *indirect* approach to inferring reproductive isolation, typically involving some measure of chromosomal or genetic differences that would preclude successful interbreeding; adaptive differences in morphology, behavior, or physiology that would reduce fitness of hybrids; and/or traits involved with mate recognition such that successful mating is unlikely between putative species. While most evolutionary biologists concur that some level of interbreeding can occur among species, a problem emerges when authors use low thresholds for evidence of reproductive isolation or these assessments are subjective (or no thresholds). A recent example of this issue is the description of the Gunnison sage grouse as a new species (Young et al. 2000), "splitting" it off from the greater sage grouse. The authors of that species description primarily relied on a poorly designed vocalization playback experiment and claimed their results were evidence of reproductive isolation. The study lacked equal sampling effort, relied on playbacks that were sped up for one species and slowed down for another, and the authors relied on a subjective post hoc interpretation of results (i.e., both "species" avoided playbacks, but avoided their own playback slightly less). In terms of genetic data, only one unique mitochondrial DNA haplotype was found in Gunnison sage grouse and it was at low frequency (0.24); all other mtDNAs were common to the greater sage grouse. There was one unique nuclear microsatellite allele to the Gunnison sage grouse (out of thirty-nine alleles) but it was found in only four individuals in one population. The authors explained away these genetic results, speculating that reproductive isolation was *so recent* that divergence in genetic markers had not *yet* occurred.

Although the body size of the Gunnison sage grouse is smaller than that reported for the greater sage grouse, Young et al. (2000) exaggerated the differences between the Gunnison sage grouse and greater sage grouse by not including intermediate populations in the comparison and using the largest sage grouse population as representative of that species as a whole. Reported differences between the Gunnison sage grouse and greater sage grouse in plumage and courtship behavior were qualitative and relied on line drawings, artist renderings, or pictures of presumably "typical" individuals. Taken collectively, the authors' claims that these results are indicative of reproductive isolation under the Biological Species Concept must be viewed with some skepticism (Ramey 2005, and in prep).

In more than one case, recent geographic separation, often due to human activities (e.g., urban or agricultural development since the early twentieth century), was used as evidence of reproductive isolation to describe "new" subspecies. One of the most significant but least known examples is the subspecies of beach mice described by Bowen in 1968. Bowen (1968) dismissed a previous author's (Sumner 1929) conclusions about finding clinal variation (gradually transitioning) in pelage color and pattern of Florida beach mice (*Peromyscus polionotus*) as "misinterpreted." However, Bowen (1968) did not provide a quantitative basis for this conclusion. Instead, Bowen (1968) suggested that several intervening and previously undescribed subspecies between the Gulf Coast barrier islands and inland populations had "confused" Sumner's interpretation of the data. In Bowen's view, subspecies could be very recent in origin. From geological studies he knew that a number of the barrier islands were less than six thousand years old and populations of mice on these islands were subject to frequent extinction events (due to hurricanes, storm surge, sea level rise, and island erosion), as well as recolonization from other islands and the mainland, thus diluting any recent genetic divergence. But most amazing are the two primary isolating mechanisms cited by Bowen (1968) in support of his "new" subspecies of beach mice: dredging of the Gulf Intracoastal Waterway and breaches in barrier islands from the 1843 hurricane. All three of Bowen's "new" subspecies: the Alabama beach mouse (*Peromyscus polionotus ammobates*), Perdido Key beach mouse (*Peromyscus polionotus tryssyllepsis*), and Choctawatchee beach mouse (*Peromyscus polionotusallophrys*) are now listed as "endangered" under the ESA.

In more recent years, even slight but statistically significant differences in allele frequencies (or minor behavioral and/or morphological variation) are considered evidence of reproductive isolation, thereby elevating many subspecies to the level of species and many populations to the level of subspecies. Such loose interpretations of the BSC run the risk of uncritically equating statistical significance with biological significance (see discussions by Isaac, Mallet, and Mace 2004; Agapow et al. 2004; Meiri and Mace 2007).

A third cause of taxonomic inflation is the desire to elevate some populations or subspecies to the level of species out of concern for their conservation status—the logic being that erring on the side of caution requires erring on the side of species status, an "ends justify the means" rationale (Chaitra, Vasudevan, and Shanker 2004; Isaac, Mallet, and Mace

2004; Meiri and Mace 2007). As noted above, the listing of taxa or populations brings regulatory actions and funding associated with one of the world's most powerful environmental laws, the ESA. However, once listed as threatened or endangered, or considered for candidate status under the ESA, the die is cast and even if compelling contrary scientific evidence is presented, it is extremely unlikely to be reversed. Environmental activists, as well as similarly motivated academics, professional societies, and agency employees have acted to oppose taxonomic revision, or failed to acknowledge contrary information, even in cases of gross taxonomic error. The prolonged and politically charged defense of the so-called Preble's meadow jumping mouse subspecies listing is an example of a listing maintained through obfuscation, intimidation, and ignoring contrary evidence by all parties mentioned above. That case is well documented in scientific literature (Crandall 2006; Matrin 2006; Ramey et al. 2006, 2007), written congressional testimony (Ramey 2007), the press (Holthouse 2005; Carrol 2006a,b; Moore 2006), and senate correspondence (see letter by Senator W. Allard of Colorado to D. Kempthorne, Department of Interior, dated 30 July 2007).

Fourth and finally, taxonomic inflation has been further exacerbated by the popular convention of referring to species, subspecies, and even populations as if they were "full" species. This practice has become commonplace as a result of a linguistic loophole in the ESA. The loophole occurs because the ESA defines "species" in an unusual way. The statute and the regulations that flow from it allow for the federal listing of species, subspecies, and DPSs as threatened or endangered; and because the statute refers to all three categories as *species*, it has become increasingly common for conservation biologists, environmental advocates, the news media, counsel, and lawmakers to refer to all of these as *species*. At best, this is a confusing practice. At worst, it fosters an erroneous perception that all such ESA-listed entities are *full* species. This perceptual problem is compounded when common names are used rather than Linnean nomenclature—one type of misunderstanding that Linnean nomenclature was intended to avoid (e.g., the California condor (*Gymnogypus californianus*) is a species; the Mexican wolf (*Canis lupus baileyi*) is one of twenty-seven subspecies of wolf described in North America; the Pacific coast population of the western snowy plover (*Charadrius alexandrinus nivosus*) is a DPS.

The plethora of species concepts and definitions, as well as a general lack of operational definitions with thresholds to test species, has meant that species descriptions are often based on subjective, post hoc interpretations of data rather than on a hypothetico-deductive approach using thresholds set in advance of data collection, that is, *the scientific method* (Popper 1962; Platt 1964). If species concepts and definitions can be selected post hoc to fit any set of observations, then just about any group of organisms could potentially qualify (or not qualify) as a species depending on the investigator's whim or regulatory agency's bias (Hull 1965; Ramey et al. 2006; Wilson and Brown 1953; Hennig 1966). Such an approach not only fosters taxonomic inflation but eventually renders *species* a meaningless term, and reliance on information derived from such an approach puts conservation and regulatory decisions on shaky epistemological grounds. Subspecies and DPSs are even more problematic (Cronin 1997, 2006, 2007; Ramey et al. 2005, 2006, 2007).

PROFESSIONAL SOCIETIES AS AUTHORITATIVE SOURCES OF TAXONOMIC INFORMATION

It may come as a surprise to outside observers, but there are no "rules" in taxonomy as to what constitutes a species (or subspecies), nor consistently applied thresholds for defining them within or across different kinds of organisms (e.g. mammals, butterflies, or vascular plants). The International Code of Zoological Nomenclature, sometimes referred to as "The Code" (International Commission on Zoological Nomenclature 1999), provides only guidelines for the process by which species are formally described. Anyone can describe "new" species or combine (synonymize) former ones; it is simply a matter of formally describing the previous taxonomic classification (Latin name, author, and references) and providing a new description and name (as well as location of any type specimens if appropriate). There are further guidelines on wide dissemination and access to the taxonomic description, and no requirement for peer review. There is no requirement that any of the underlying data relied upon be made publicly available. While most taxonomic revisions are presently published in peer-reviewed journals, the quality of peer review varies widely among journals. And because peer reviews are typically provided free, as a professional courtesy, few peer reviewers take the time to reanalyze data, even if the data are available.

One would expect that professional societies would provide a level of objectivity, rigor, and consistency in taxonomic decisions. However, this is not necessarily the case. For example, the American Society of Mammalogists has no standing committee by which to judge taxonomic decisions and no formal process for accepting or rejecting species from its officially recognized list of "Mammal Species of the World." Yet, despite these limitations, and the fact that the list is updated infrequently (the last update was in 2005), this is cited as authoritative work. Furthermore, professional societies are not necessarily neutral authorities or uninterested parties in taxonomic decisions involving the ESA. For example, a former president of the American Society of Mammalogists (ASM) went on record as opposing the taxonomic revision that would have led to the delisting of the so-called Preble's meadow jumping mouse. The letter was drafted by activists within the ASM (letter from G. Cameron, ASM to S.

Wiley, USFWS, dated 26 April 2005; ASM Conservation Committee correspondence 2005).

Similarly, the American Ornithologists Union (AOU) sanctioned the recognition of the coastal California gnatcatcher as a subspecies and advocated for its listing under the ESA. The AOU's Committee on Classification and Nomenclature (CCN) is the official body that reviews proposals for taxonomic revision but the process falls far short of the peer review process used by many academic journals. Proposals are reviewed based on limited information and must receive a two-thirds positive vote by committee members to pass. The shallow depth of comment on recent taxonomic proposals, often less than a paragraph, may be viewed at the AOU-CCN's website, www.aou.org/committees/nacc/proposals/pending.php. In the case of older taxonomic decisions, such as the recognition of the Gunnison sage grouse as a new species, the records of AOU-CCN comments or deliberations may no longer exist. The following comments by a former chairman of that committee illustrate the problem with decisions being based on predetermined conclusions rather than on a rigorous and independent peer review: "Frankly, I doubt that there were many comments. It was a clear-cut new species, and we had all known about it for some time. There was not much to question—except whether the description was acceptable. I checked with 'experts' on that and it seemed valid." (e-mail from R. Banks to R. Ramey, dated 30 May 2008).

In summary, there are no opportunities for public comment or public deliberations on taxonomic decisions by professional societies, and no requirement for independent peer review. Yet the USFWS has relied upon the AOU, ASM, and other societies as an authoritative source of taxonomic information. The shortcomings and potentially partial nature of taxonomic decisions by professional societies raises legal questions regarding the reliance of federal agencies on information provided by them.

PEER REVIEW ISSUES

In an ideal world free of conflict of interest or personal agendas, peer review provides an essential filter on the quality of information used in science and decision making. This vetting process can provide opportunities for constructive feedback, identifying shortcomings in the research, and a chance to replicate results. Research that is seriously flawed or overstates results is winnowed out. While the peer reviews used in most scientific journals fit this model, and policies are in place to ensure it, the independence and objectivity of peer reviews can be compromised if there is a conflict of interest or personal agendas influencing a reviewer's opinions. Peer review can also miss errors in analyses, especially if the data used in analyses are not provided to reviewers, as reviewers must then assume its accuracy.

The USFWS regularly utilizes peer reviews in cases of contentious taxonomic reassignments that could affect listing status (peer reviews are also used for proposed listings, critical habitat designations, and other agency actions). Despite the best of intentions, these peer reviews are not necessarily free of conflict of interest, nor do they always deliver an impartial assessment. The problems that lead to these issues are threefold. First, the number of experts involved with a particular species is often limited. Whole careers are sometimes dedicated to the study of a species (or subspecies or population), and a listing can produce what is perceived as needed "protection" for that species under the ESA. Additionally, ESA listings can have the effect of putting these experts into positions of power, money, and authority, through their roles on Recovery Teams, Habitat Conservation Plans, and consulting as USFWS "approved biologists." Because few ESA-listed species are ever delisted, this guarantees a virtual lifetime of employment on one's favorite species. Thus, experts used in peer review may also be advocates, or have an emotional or financial stake in the proposed listing. (Note the previous example of peer review of the coastal California gnatcatcher taxonomy.)

Second, a network of individuals who work on a particular species (or issues common to several species) can form powerful "species cartels." These social networks can influence the peer review process, provide a united front to advocate for particular decisions, and repress the publication of information that does not agree with their positions. (This issue is well documented in the case of the so-called Preble's meadow jumping mouse.) Similar issues have arisen over designation of Critical Habitat (Turner et al. 2004, 2006).

And third, the use of other federal biologists in peer review, especially those from the USFWS and the USGS-Biological Resources Division (USGS-BRD), cannot be viewed as conflict free. As noted on the USGS website, the USGS-BRD has a close working relation with the USFWS on the science used in ESA decisions:

> We [the USGS-BRD] support the Department of the Interior agencies that manage over 500 million acres of land and waters, or about one-fifth of the land in the United States . . . To meet the science needs of these agencies, as well as States and other private and public organizations, the USGS conducts investigations on species listed under the Endangered Species Act of 1973 (ESA), the Marine Mammal Protection Act, and other Federal and State laws and regulations. Working with our partners, we are focusing research on efforts to restore species listed under ESA. USGS studies how ecosystems and species interact to find ways to keep ecosystems functioning and species populations healthy. We are also conducting research on "candidate" species, that

is, species that are thought to be at risk but have not yet been listed as threatened or endangered under ESA.

The increasing codependency of the USFWS and USGS-BRD results in a growing and previously unrecognized conflict of interest in science used in support of ESA decisions and the use of USGS biologists as peer reviewers on information used in ESA decisions. This extends to the role of USGS biologists who serve as editors and reviewers for scientific journals, and who peer review highly influential scientific information used in ESA decisions.

The case of the coastal California gnatcatcher illustrates how far the peer review can diverge from accepted practice when the reviewers are from the same federal agency as the one whose listing decision is being questioned.

In October of 2004, the USFWS initiated a review of California gnatcatcher subspecies taxonomy. The USFWS stated that the review panel was to be an "objective peer review," however, because it included five USFWS and two other federal biologists it cannot be considered impartial. The use of USFWS biologists (or federal biologists in general) for a USFWS peer review of the information used in a USFWS decision clearly represents a conflict of interest.

The information provided to participants was limited, and only included reports or papers, not the data underlying them. This included the original paper describing the coastal California gnatcatcher as a subspecies (Atwood 1991), a paper in which the authors further split the coastal California gnatcatcher into two subspecies (Mellink and Rea 1994), a theoretical paper proposing a 75 percent rule for diagnosing subspecies (Patten and Unitt 2002), and a paper that found no support for the subspecies designation based on analysis of mtDNA sequences (Zink et al. 2000). The reviewers were also provided with copies of the peer reviews on a Proposed Rule to designate critical habitat for the Coastal California gnatcatcher and potential determination of a DPS for the California gnatcatcher (a DPS designation is sometimes used as a fall-back position if a subspecies should be found to be invalid). The reviewers then participated in a review process that was called "The Coastal California Gnatcatcher Taxomony Exercise" which was intended to inform senior decision makers at the Department of Interior.

The information provided to peer reviewers, however, was incomplete and did not include a substantial body of information that did not support the recognition of the subspecies. The reviewers were not provided with Atwood's original data (which Atwood did not release until it was obtained under court order earlier) nor first critical review of Atwood's 1988 monograph by Barrowclough (1992), nor technical reports or draft papers by statisticians and systematists that extensively retested Atwood's results using his now publicly available final data set (Barrowclough 1994; Brush 1994; Grant 1994; Link and Pendelton 1994; McDonald et al. 1994; Messer 1994; Boyce 1995; Cronin 1995; Roberts and Bayn 1994; Skalski 1995a, b). Also not included were admissions by Atwood in a deposition that he had "disposed of" his raw data forms before finishing his dissertation; that he "may have used error corrections" (estimates derived from pooled population means rather than actual measurement data) to replace missing values in the data set used in his analyses but "did not know if it was done or not"; and used "subjective criteria of my own" to include or exclude specimens but did not keep any record of those determinations, and had "serious doubts" about the use of spectrophotometer data used to quantify color measurements (variables critical to recognition of the subspecies) (letter from R. Thornton, Nossaman to G. Kobetich, USFWS, dated Dec. 1, 1994, with Atwood's (1994) deposition transcript included as an exhibit.).

The reanalyses of Atwood's data found no support for a "break" in morphological characters at 30° N latitude, as claimed by Atwood. Instead, morphological variation was found to be part of a north-south cline. Despite these contrary results and Atwood's admissions, the USFWS explained away these issues and reaffirmed its recognition and listing of the coastal California gnatcatcher subspecies (U.S. Fish and Wildlife Service 1995). Two more recently published papers, including one based on an extensive reanalysis of Atwood's data, brought up additional issues which further refute the basis of this specious subspecies (Cronin 1997; Skalski et al. 2008). The participants of the Gnatcatcher Taxonomy Exercise subsequently reported in a Powerpoint presentation to decision makers that: "Significant scientific controversy continues to exist and is not resolved." Also, "the morphological variation indicates geographic variation." And, "Mitochondrial DNA analysis alone is not enough to revise current taxonomy." In their final slide, the group recommended options for the coastal California gnatcatcher including convening an "independent scientific body to reanalyze morphological data and collect additional (nuclear) genetic data." This final recommendation illustrates three key issues. First, if reviewers are not provided a full package of information, or organizers have been selective in the information provided, they can come to erroneous conclusions. The reviewers could have been simply unaware of the extensive reanalysis of Atwood's data or how to analyze the data. Second, while additional evidence is almost always called for, critical tests that could potentially falsify the claim that a subspecies is unique are typically lacking. In this case, reviewers called for nuclear genetic data from microsatellites, but they did not provide any thresholds with which they would accept or reject a subspecies or DPS based on these data alone. And third, there was no questioning of the extent to which morphological variation (in this case, plumage or color) could be under environmental rather than genetic influence. The

ESA (and DPS policy) does not provide for the listing of subspecies or DPS on the basis of environmental differences alone.

The solutions to the issues identified with peer review will require a concerted effort by the USFWS to adopt procedures that will ensure independent reviews and hold the agency and staff accountable for those reviews. While previous proposals have been made to improve the peer review used in ESA decisions (Fleishman 2001; Hecht and Parkin 2001), they do not address the issues raised here. A useful start would be to adopt the same conflict of interest policy as used by the National Academy of Sciences to minimize potential for financial and other conflicts of interest. Internal agency peer reviews should be strongly discouraged. The USFWS need not abrogate its responsibilities by hiring outside contactors, it simply needs the policies in place to administer peer reviews in an objective and consistent manner. Next, the USFWS could take steps to ensure that all information, including contrary information, is fully considered in peer reviews of listing/delisting decisions. A consistent use of more narrowly defined questions than asked at present would serve conservation by avoiding the responses that are simply opinion surveys.

DATA ACCESS AND CONTROL

Presently, the USFWS does not require the data used in peer-reviewed research to be made publicly available; it relies only on the published studies. However, a requirement for the data to be made publicly available (perhaps in an online archive) would raise the bar on the quality and repeatability of science used in ESA taxonomic decisions. This would also eliminate problems associated with trying to obtain data from individual researchers (especially problematic after they pass away). Overall, this would not be a burdensome requirement, as an increasing number of scientific journals require that data used in papers be archived. It is also consistent with a requirement of the Information Quality Act (Section 515 of the Treasury and General Government Appropriations Act for Fiscal Year 2001, Pub. L. No. 106-554; H.R. 5658) and a recently issued Scientific Integrity Memorandum by the President (Obama 2009).

AGENCY DEFERENCE

Disputable taxonomic classifications are anathema to governments that prefer some level of scientific certainty prior to committing to difficult regulatory decisions. A lack of consensus on operational definitions, the absence of consistently applied quantitative thresholds, data sets that were never made public, and the continued use of antiquated taxonomic descriptions all contribute to this growing problem (Cronin 1997, 2006, 2007; Fischman and Meretsky 2001; Quarles 2001). As a result, governmental agencies such as the U.S. Fish and Wildlife Service (USFWS) and National Marine Fisheries Service rely on the opinions of taxonomists, peer-review committees, interest groups, professional societies, and internal agency reviewers to determine if a taxon is valid. The problem with such reviews is that they are not held to any threshold of uniqueness or standards of evidence that could be consistently used to accept or reject a species, subspecies, or DPS (Ramey 2007). Because of that, even peer reviews can become de facto opinion surveys or be based on less than a complete record, leaving species very much "in the eye of the beholder." Furthermore, in cases of disputable species, recommendations for additional studies that are made without recommendations for thresholds to be used in a hypothesis testing framework only serve to perpetuate the problem. Using opinion rather than falsification as a basis for resolving taxonomic disputes has the implication of effectively putting these decisions outside the realm of science (Popper 1962).

While such an approach may lead to disputable taxonomic decisions with far-reaching regulatory implications, the discretion of the USFWS is generally shielded by so-called Chevron deference—a legal precedent that requires the courts to defer to agency interpretations of ambiguous statutory provisions, such as the terms "species," "subspecies," and "DPS" as used in the ESA. (*Chevron U.S.A., Inc. v. Natural Resources Defense Council, Inc.*, from 1984, and *United States v. Mead Corp.*, from 2001). The USFWS may insist that Congress has delegated to it the job of giving precise meaning to these terms, relying upon that agency's expertise. However, given the multiplicity of scientific approaches to taxonomic categorization, it is hardly surprising that the USFWS has applied inconsistent criteria to the recognition of subspecies and DPSs. In the words of one endangered species legal expert testifying before the U.S. Senate on the ambiguous language used to define DPSs: "*With this amount of discretion [by the USFWS], the eyesight of the beholder can be quite poor, and yet suffice*" (Quarles 2001).

POTENTIAL APPROACHES TO SETTING CONSERVATION PRIORITIES

1. Minimum Thresholds

While it may not be possible to develop thresholds that can be applied across all species or subspecies, it would be possible to arrive at standards within major groups (e.g. mammals). The primary caveat is analogous to type I and II statistical error: If the bar is set too high, then some unique organisms may fail to be recognized; yet, if the bar is set too low, then any population could potentially qualify as a species, contributing to taxonomic inflation. The criteria themselves are subject to debate, but the need for explicit criteria is obvious (Sites and Marshall 2003; Ramey et al. 2005; Crandall 2006). While appropriate thresholds can be debated and revised, the first step in establishing standards is to state them explicitly. If consistent criteria were to be set in advance of data collection, this hypothesis-testing approach would transform a subjective system of classification into an objective one.

We used quantitative criteria for testing subspecies and DPS validity by employing an explicit minimum thresholds approach with multiple data sets, complete spatial sampling, and criteria set in advance of data collection (Ramey et al. 2005, 2006, 2007). Each of these used threshold levels for various tests that have some conventional history below the level of species (e.g. Worley et al.'s (2004) use of a probability greater than 0.90 as a standard in assignment tests, Wehausen and Ramey's (2000) use of > 90 percent correct assignment using posterior probabilities of $P > 0.95$ in linear discriminant analysis for morphometric data). The goal was to establish reasonable threshold levels for these sorts of tests where they have often been absent, adapting quantitative criteria from the conceptual approaches of Avise and Ball (1990), Ball and Avise (1992), and Crandall et al. (2000). The basic idea is that congruence among multiple data sets suggests a higher degree of confidence in genetic uniqueness and thus a higher conservation priority.

2. *Phylogenetic Ranking*

Another way of avoiding the species debate is to prioritize the rank of groups regardless of what they are called (e.g., species, subspecies, DPS) based on their degree of evolutionary distinctiveness. The Zoological Society of London (ZSL) has developed a quantitative, phylogenetic approach to ranking species and their conservation priorities based on measures of evolutionary distinctiveness and level of endangerment (Hendry et al. 2000; Isaac et al. 2007; see ZSL website www.edgeofexistence.org). This method, dubbed "Evolutionarily Distinct and Globally Endangered" (EDGE), gives higher rankings to species that are monotypic genera (a single species in the genus) and lower rankings to species that are one of many found in a genus. For example, in ranking 4,173 mammal species, the top three evolutionarily distinct species were found to be the duck-billed platypus (*Ornithorhynchus anatinus*), aardvark (*Orycteropus afer*), and short-beaked echidna (*Tachyglossus aculeatus*). Clearly, all of these represent evolutionarily distinct lineages. In contrast, speciose groups that are widespread, such as the murid rodents (mice), tend to rank low on the list of conservation priorities because their lineages are well represented.

This approach is similar in principle to the temporal ranking scheme first proposed by Avise and Johns (1999). The attractiveness of the EDGE approach (which is based on a hierarchical ranking using phylogenetic measures of uniqueness and degree of extinction threat) is that it solves some of the inherent problems associated with drawing lines through a continuum of variation. Groups are ranked across the entire range of genetic variation in major taxonomic affinities (e.g. mammals, birds, vascular plants). It also provides an objective means for setting conservation priorities and therefore some degree of regulatory consistency under biodiversity laws and treaties. If this approach were expanded to include species and presumably distinct populations, setting conservation priorities would acquire a level of scientific rigor and repeatability that is presently lacking. Such a ranking system is robust and can readily adapt to the introduction of new information (Isaac et al. 2004).

3. Conservation Triage

A third approach would involve "Conservation Triage," a strategy analogous to medical triage, and defined as: "*Triage in a conservation context is the process of prioritizing the allocation of limited resources to maximize conservation returns, relative to the conservation goals, under a constrained budget*" (Bottrill et al. 2008). As noted by the authors, in an era of budgetary constraints and an increasing number of ESA listings: "*Decision making based on the principles of triage provides a defensible, rational and repeatable approach to prioritizing conservation investments.*" Although the USFWS currently has a ranking system, based on the 1982 amendments to the ESA, there is no correlation between the priority ranking and the funding allocation. The current priority ranking system is also extremely crude: "species" uniqueness is ranked on three levels: monotypic genus, species, and subspecies/DPS. And while the USFWS tracks their agency's ESA expenditures, there is no formal accounting for conservation expenditures by other federal and state agencies, local governments, NGOs, or private parties, outside of the economic analyses that accompany Critical Habitat designations. And finally, the USFWS evaluates success of recovery programs only in vague terms: stable, increasing, or decreasing. Therefore, the approach proposed by Bottrill et al. (2008), which employs quantitative measures of phlyogenetic distinctiveness (see discussion of EDGE above), biodiversity benefit, probability of success, and cost, into their priority ranking, could be adapted for use in ESA decision making, while largely avoiding the issue of how to best define species, subspecies, and DPSs.

THE PATH AHEAD

Would implementation of the ESA's conservation efforts be more effective if species, subspecies, and populations were tested against consistent threshold(s) of genetic uniqueness prior to being listed as endangered?

Or would a ranking of taxa and populations based on phylogenetic uniqueness similar to the EDGE program provide a more workable approach? Could we embark upon the paradigm shifts required to implement conservation triage? Any of the three approaches described above could provide a more consistent basis for allocation of conservation resources to evolutionarily distinct taxa while winnowing out erroneously designated ones. Clearly, there are now more precise conceptual and analytical tools that can be used to objectively distinguish or rank phylogenetic distinctiveness of species, subspecies, and distinct population segments than there were in the early twentieth century (e.g., Crandall et al. 2000; Sites and Marshall 2003; Baker and Bradley 2006). In addition, the application of a consistent method for prioritizing conservation effort could remove much of the subjectivity that has dominated taxonomic decisions over the last several hundred years, and consequently hindered effective implementation of the ESA. Until such a paradigm shift takes place, or is forced onto the system, species will likely remain "in the eye of the beholder."

With twenty-seven definitions and a long-running disagreement over "Just what is a species?" (Isaac, Mace, and Mallet 2005; Hey 2006; Mallet 2007a, 2007b), it does not seem likely that taxonomists will be the ones to ultimately decide the issue for the rest of society. Nor is it likely that governmental agencies, executive branches, or legislative branches will decide, for similar reasons. Therefore, it will ultimately be the courts that decide what constitutes a "species," simply because they are the ultimate arbiter in matters of law and public policy. Whether the courts take on the issue and can arrive at a biologically meaningful and legally defensible definition of species, or abandon its use entirely in favor of a phylogenetic ranking system or other approach, is open to question. The outcome will depend, in part, upon our collective abilities to break with the subjective approach of the past and identify well-reasoned and objective alternatives for the future.

REFERENCES

Agapow, P. M., O. R. P. Bininda-Emonds, K. A. Crandall, J. L. Gittleman, G. M. Mace, J. C. Marshall, and A. Purvis (2004). "The Impact of Species Concept on Biodiversity Studies." *Quarterly Review of Biology* 79(2): 161–179.

ASM Conservation Committee (2005). List server correspondence found via Google internet search using keywords "Ramey" and ".txt" lighthouse.valdosta.edu/pipermail/asmcc/attachments/20050428/96c072b5/prebleiletter042605-0001.doc, lighthouse.valdosta.edu/pipermail/asmcc/2005-February.txt, lighthouse.valdosta.edu/pipermail/asmcc/2005-March.txt, lighthouse.valdosta.edu/pipermail/asmcc/2005-April.txt accessed April 1–May 1 2005, copies of correspondence available from the author.

Atwood, J. L. (1991). "Subspecies Limits and Geographic Patterns of Morphological Variation in California Gnatcatchers (*Polioptila californica*)." *Bulletin of the Southern California Academy of Sciences* 90(3): 118–33.

Atwood, J. L. (1994). "Endangered Species Committee vs. Babbitt et al.," Miniscript Depositions, Jonathan Lee Atwood, Volumes I, II, and III, Taken 15–17 November 1994. Leavitt & Finstein Reporting Service, Wemouth, MA.

Avise, J. C., and R. M. Ball Jr. (1990). "Principles of Genealogical Concordance in Species Concepts and Biological Taxonomy." In *Oxford Surveys in Evolutionary Biology*, Vol. 7: *1990*, edited by Douglas Futuyma and Janis Antonovics. Oxford: Oxford University Press.

Avise, J. C., and G. C. Johns (1999). "Proposal for a Standardized Temporal Scheme of Biological Classification for Extant Species." *Proceedings of the National Academy of Sciences of the United States of America* 96(13): 7358–7363.

Baker, R. J., and R. D. Bradley (2006). "Speciation in Mammals and the Genetic Species Concept." *Journal of Mammalogy* 87(4): 643–62.

Ball, R. M., Jr., and J. C. Avise (1992). "Mitochondrial DNA Phylogeographic Differentiation among Avian Populations and the Evolutionary Significance of Subspecies." *Auk* 109(3): 626–36.

Barrowclough, G. F. (1992). "Speciation and Geographic Variation in Blacktailed Gnatcatchers," (Book review of Jonathan L. Atwood. 1988. Ornithological Monograph no. 42. American Ornithologists' Union, Washington, DC. 74 p. ISBN 0-943610-53-2.) *Condor* 94: 555–556.

Barrowclough, G. F. (1994). "Comments on the Subspecific Taxonomy of California Gnatcatchers Based on an Analysis of a Statistical Study by McDonald et al." Unpublished report, New York.

Bottrill, M. C., L. N. Joseph, J. Carwardine, M. Bode, C. Cook, E. T. Game, H. Grantham, S. Kark, S. Linke, E. McDonald-Madden, R. L. Pressey, S. Walker, K. A. Wilson, and H. P. Possingham (2008). "Is Conservation Triage just Smart Decision Making?" *Trends in Ecology and Evolution* 23(12): 649–54.

Bowen, W. W. (1968). "Variation and Evolution of Gulf Coast Populations of Beach Mice, *Peromyscus polionotus*." *Bulletin of the Florida State Museum* 12:1-91.

Boyce, M. S. (1995). "Quantitative Basis for Subspecies Delineation of the California Gnatcatcher (*Polioptila californica*): A Review." Unpublished report. College of Natural Resources, University of Wisconsin.

Brush, A. H. (1994). "Taxonomic Status of the California Gnatcatchers of Northwestern Baja California, Mexico." Comments by Alan H. Brush. Unpublished report. Logan, UT.

Carrol, V. (2006a). "On Point: The Mouse that Roared." *Rocky Mountain News* April 5, 2006. Denver, CO.

Carrol, V. (2006b). "On Point: Species Act Abused." *Rocky Mountain News.* April 6, 2006. Denver, CO.

Chaitra, M. S., K. Vasudevan; and K. Shanker (2004). "The Biodiversity Bandwagon: The Splitters Have It." *Current Science* 86(7): 897–99.

Chevron U.S.A., Inc. v. Natural Resources Defense Council, Inc., 467 U.S. 837 (1984).

Crandall, K. A. (2006). "Advocacy Dressed Up as Scientific Critique." *Animal Conservation* 9 (3): 250–51.

Crandall, K. A., O. R. P. Bininda-Emonds, G. M. Mace, and R. K. Wayne. (2000). "Considering Evolutionary Processes in Conservation Biology." *Trends in Ecology and Evolution* 15(7): 290–295.

Cronin, M. A. (1993). "Mitochondrial DNA in Wildlife Taxonomy and Conservation Biology: Cautionary Notes." *Wildlife Society Bulletin* 21(3): 339–48.

Cronin, M. A. (1995). "Comments Regarding California Gnatcatcher (*Polioptila californica*) Taxonomy and Population Management Units." Unpublished report to the U.S. Fish and Wildlife Service. LGL Alaska Research Associates, Inc., Anchorage, Alaska.

Cronin, M. A. (1997). "Systematics, Taxonomy, and the Endangered Species Act: The Example of the California Gnatcatcher." *Wildlife Society Bulletin* 25 (3): 661–66.

Cronin, M. A. (2006). "A Proposal to Eliminate Redundant Terminology for Intraspecies Groups." *Wildlife Society Bulletin* 34(1): 237–41.

Cronin, M. A. (2007). "The Preble's Meadow Jumping Mouse: Subjective Subspecies, Advocacy, and Management." *Animal Conservation* 10(2): 159–61.
Darwin, C. (1859). *On the Origin of Species by Means of Natural Selection; or, The Preservation of Favoured Races in the Struggle for Life*. London: John Murray.
Fischman, R. L., and V. J. Meretsky. (2001). "Endangered Species Information: Access and Control." *Washburn Law Journal* 41(1): 90–113.
Fleishman, E. (2001). "Moving Scientific Review Beyond Academia." *Conservation Biology* 15 (3): 547–49.
Grant, C. V. (1994). "Critique of Atwood's (1991) use of data, the latitudinal density of the California gnatcatcher in baja California and recommended action for Fish and Wildlife Service." Unpublished report. Bio-Resources, Inc., Logan, Utah.
Hall, E. R., and K. R. Kelson. (1959). *The Mammals of North America*, vol. 2. New York: Ronald Press.
Hecht, A., and M. J. Parkin. (2001). "Improving Peer Review of Listings and Recovery Plans under the Endangered Species Act." *Conservation Biology* 15(5):1269–1273.
Hendry, A. P., S. M. Vamosi, S. J. Latham, J. C. Heibuth, and T. Day (2000). "Questioning Species Realities." *Conservation Genetics* 1(1): 67–76.
Hennig, W. (1966). *Phylogenetic Systematics*, translated by D. Dwight Davis and Rainer Zangerl. Urbana: University of Illinois Press.
Hey, J. (2006). "On the Failure of Modern Species Concepts." *Trends in Ecology and Evolution* 21(8): 447–50.
Holthouse, D. (2005). "Building a Better Mousetrap: Biologist Rob Ramey isn't Afraid of Taking Risks—but Taking on Environmentalists may be his Riskiest Move." *Westword* published 1 January 2005. Denver, CO.
Hull, D. L. (1965). "The Effect of Essentialism on Taxonomy: Two Thousand Years of Stasis." *British Journal for the Philosophy of Science* 15(60): 314–26.
International Commission on Zoological Nomenclature (ICZN) (1999). *International Code of Zoological Nomenclature* 4th ed., London: ICZN.
Isaac, N. J. B., G. M. Mace, and J. Mallet (2005). "Response to Agapow and Sluys: The Reality of Taxonomic Change." *Trends in Ecology and Evolution* 20(6): 280–81.
Isaac, N. J. B., J. Mallet, and G. M. Mace (2004). "Taxonomic Inflation: Its Influence on Macroecology and Conservation." *Trends in Ecology and Evolution* 19(9): 464–69.
Isaac, N. J. B., S. T. Turvey, B. Collen, C. Waternan, and J. E. M. Ballie (2007). "Mammals on the EDGE: Conservation Priorities Based on Threat and Phylogeny." *PLoS ONE* 2(3): e296. Available from www.plosone.org/article/fetchArticle.action?articleURI=info%3Adoi%2F10.1371%2Fjournal.pone.0000296.
King, T. L., J. F. Switzer, C. L. Morrison, M. S. Eackles, C. C. Young, B. A. Lubinski, and P. Cryan (2006). "Comprehensive Genetic Analyses Reveal Evolutionarily Distinction of a Mouse (Zapus hudsonius preblei) Proposed for Delisting from the U.S. Endangered Species Act." *Molecular Ecology* 15: 4331–4359
Link, W. A., and G. W. Pendelton (1994). "New Analyses of Atwood's California Gnatcatcher Data." Unpublished report.
Linnaeus, C. (1735). "Systema naturae per regna tria naturae: secundum classes, ordines, genera, species, cum characteribus, differentiis, synonymis, locis."
Linnaeus, H. (2007). "Conservationists—and Polar Bears—Should Heed the Lessons of Economics." *The Economist* 383(8529): 13.
Mallet, J. (2007a). "Species, Concepts of." In *Encyclopedia of Biodiversity* vol. 5, 2nd ed. Edited by Simon Asher Levin. San Diego, CA: Academic Press.
Mallet, J. (2007b). "Subspecies, Semispecies, Superspecies." In *Encyclopedia of Biodiversity* vol. 5, 2nd ed. Edited by Simon Asher Levin. San Diego, CA: Academic Press.
Marris, E. (2007). "The Species and the Specious." *Nature* 446 (15): 250–53.
Martin, A. (2006). "Advocacy Dressed Up As Science: Response to Ramey et al." *Animal Conservation* 9:248–49.
Mayr, E. (1970). *Populations, Species, and Evolution*. Cambridge, MA: Harvard University Press, Belknap Press.

McDonald, L. L., W. P. Erickson, and J. W. Kern (1994). "Statistical Results Concerning Subspecies Limits and Geographic Patterns of Morphological Variation of California Gnatcatchers (*Polioptilacalifornica*)." Based on data provided by Dr. Johathan L. Atwood, 19 May 1994. Unpublished report by WEST, Inc., Cheyene, WY.

Meiri, S., and G. M. Mace (2007). "New Taxonomy and the Origin of Species." *PLoS Biology* 5(7): 1385–1386.

Mellink, E., and A. M. Rea (1994). "Taxonomic status of the California gnatcatchers in northwestern Baja California." *Western Birds* 25: 50–62.

Merriam, C. H. (1918). *Review of the Grizzly and Big Brown Bears of North America (Genus* Ursus*) with the Description of a New Genus,* Vetularctos. North American Fauna 41. Washington, DC: U.S. Government Printing Office.

Messer, K. (1994). "Comments for the Public Record on the Status of the Coastal California Gnatcatcher." Unpublished report. Department of Mathematics, California State University, Fullerton, CA.

Moore, S. (2006). "Of Mice and Men: A Tiny Rodent is the Hottest Political Issue in Colorado." *Wall Street Journal,* Editorial Page. Published 23 March 2006.

Obama, B. H. (2009). "Scientific Integrity Memorandum for the Heads of Executive Departments and Agencies." Presidential Documents. *Federal Register* 74(46): 10671-10672.

Patten, M. A., and P. Unitt (2002). "Diagnosability Versus Mean Differences of Sage Sparrow Subspecies." *Auk* 119: 26–35.

Platt, J. R. (1964). "Strong Inference: Certain Systematic Methods of Scientific Thinking May Produce Much More Rapid Progress than Others." *Science* 146(3642): 347–53.

Popper, K. R. (1962). "Science: Conjectures and Refutations." In *Conjectures and Refutations: The Growth of Scientific Knowledge.* New York: Basic Books.

Quarles, S. P. (2001). "Issues Concerning the Listing of Distinct Population Segments of Vertebrates under the Endangered Species Act of 1973." Testimony before the Subcommittee on Fish, Wildlife, and Water, Environment, and Public Works Committee, U.S. Senate, May 9, 2001. Available from epw.senate.gov/107th/qua_0509.htm.

Ramey, R. R., II (2005). "Taxonomy of the Coastal California Gnatcatcher: A Review of Current Information." Unpublished report to Department of Interior. 31 March 2006.

Ramey, R. R., II (2007). "Written Testimony before the Committee on Natural Resources," U.S. House of Representatives, for the hearing *Crisis of Confidence: The Political Influence of the Bush Administration on Agency Science and Decision-Making.* 110th Cong., 1st sess., July 31, 2007.

Ramey, R. R., II, H.-P. Liu, C. W. Epps, L. Carpenter, and J. D. Wehausen (2005). "Genetic Relatedness of the Preble's Meadow Jumping Mouse (*Zapus hudsonius preblei*) to Nearby Subspecies of *Z. hudsonius* as Inferred from Variation in Cranial Morphology, Mitochondrial DNA, and Microsatellite DNA: Implications for Taxonomy and Conservation." *Animal Conservation* 8(3): 329–346.

Ramey, R. R., II, J. D. Wehausen, H.-P. Liu, C. W. Epps, and L. Carpenter (2006). Response to Vignieri et al. (2006): "Should Hypothesis Testing or Selective Post Hoc Interpretation of Results Guide the Allocation of Conservation Effort?" *Animal Conservation* 9(3): 244–47.

Ramey, R. R., II, J. D. Wehausen, H.-P. Liu, C. W. Epps, and L. Carpenter (2007). "How King et al. (2006) Define an 'Evolutionary Distinction' of a Mouse Subspecies: A Response." *Molecular Ecology* 16(17): 3518–3521.

Roberts, N., and R. L. Bayn (1994). "Replicating Atwood (1991) using data submitted to the U. S. Fish and Wildlife Service." Unpublished report. Computer Services, Utah State University, Logan, UT.

Sites, Jack W., Jr., and Jonathon C. Marshall (2003). "Delimiting Species: A Renaissance Issue in Systematic Biology." *Trends in Ecology and Evolution* 18(9): 462–70.

Skalski, J. R. (1995a). "A Cross Report Comparison of Statistical Findings Regarding the California Gnatcatcher." Unpublished report. Center for Quantitative Science, University of Washington, Seattle, WA.

Skalski, J. R. (1995b). "Critique of the Link and Pendleton (1994) Manuscript." Unpublished report. Center for Quantitative Science, University of Washington, Seattle, WA.

Skalski, J. R., R. L. Townsend, L. L. Mcdonald, J. W. Kern, and J. J. Millspaugh (2008). "Type I Errors Linked to Faulty Statistical Analyses of Endangered Subspecies Classifications." *Journal of Agricultural, Biological, and Environmental Statistics* 13(2): 199–220.

Sumner, F. B. (1929). "An Analysis of a Concrete Case of Intergradation between Two Subspecies." *Proceedings of the National Academy of Science* 15: 110-120, 481-493.

Timm, R. M., R. R. Ramey, and the Nomenclature Committee of the American Society of Mammalogists (2005). "What Constitutes a Proper Description?" *Science* 309: 2163-2164.

Turner, J. C., C. L. Douglas, C. R. Hallum, P. R. Krausman, and R. R. Ramey (2004). "Determination of critical habitat for the endangered Nelson's bighorn sheep in southern California." *Wildlife Society Bulletin*. 32 (2):427–48.

Turner, J. C., C. L. Douglas, C. R. Hallum, P. R. Krausman, and R. R. Ramey (2006). "Osterman's Assumption of a Flawed Habitat Model is Premised on Facts Not in Evidence: Turner et al. (2005) response to Osterman et al. (2005)." *Wildlife Society Bulletin* 33(4): 1465–1473.

U.S. Fish and Wildlife Service (1995). "Notice of Determination To Retain the Threatened Status for the Coastal California Gnatcatcher Under the Endangered Species Act." *The Federal Register* 60(58): 15693-15699.

United States v. Mead Corp., 121 S. Ct. 2164 (2001).

Wehausen, J. D., and R. R. Ramey II (2000). "Cranial Morphometric and Evolutionary Relationships in the Northern Range of *Ovis canadensis*. *Journal of Mammalogy* 81 (1): 145–61.

Wilkins, J. S. (2007). "Species." ScienceBlogs. Available from scienceblogs.com/evolvingthoughts/2007/01/species.php.

Wilson, E. O., and W. L. Brown Jr. (1953). "The Subspecies Concept and Its Taxonomic Applications." *Systematic Zoology* 2(3): 97–111.

Worley, K., C. Strobeck, S. Arthur, J. Carey, H. Schwantje, A. Veitch and D. W. Coltman (2004). "Population Genetic Structure of North American Thinhorn Sheep (*Ovis dalli*)." *Molecular Ecology* 13(9): 2545–2556.

Young, J. R., C. E. Braun, S. J. Oyler-McCance, J. W. Hupp, and T. W. Quinn (2000). "A New Species of Sage-Grouse from Southwestern Colorado." *Wilson Bulletin* 112: 445–53.

Zink, R. M., G. F. Barrowclough, J. L. Atwood, and R. C. Blackwell (2000). "Genetics, Taxonomy and Conservation of the Threatened California Gnatcatcher." *Conservation Biology* 14: 1394–1405.

Zink, R. M. (2004). "The Role of Subspecies in Obscuring Avian Biological Diversity and Misleading Conservation Policy." *Proceedings of the Royal Society* B 271(1539): 561–564.

Zoological Society of London (ZSL) (2012). "EDGE of Existence: Evolutionarily Distinct and Globally Endangered." Available from www.edgeofexistence.org/.

FIVE

Politics and Science in Endangered Species Act Listing Decisions

Katrina Miriam Wyman

INTRODUCTION

In April 2011 Congress crossed a line: for the first time, it delisted a species under the Endangered Species Act (the ESA or the Act) (Department of Defense and Full-Year Continuing Appropriations Act of 2011, P.L. 112-10, §1713; CRS 2011, 8). Gray wolves in the Rocky Mountain states (except Wyoming) have the honor of being the first species to lose the Act's protections through Congressional action. Congress's delisting came after a series of efforts, starting in the early 2000s, by the Fish and Wildlife Service (FWS) to downlist and delist wolves had been blocked by the courts at the behest of environmental groups. Thus the Congressional delisting achieved the agency's longstanding goal of delisting many Rocky Mountain wolves. This is evident from the fact that Congress implemented the delisting by requiring the reissuance of an FWS rule delisting many of the wolves that had been judicially vacated and insulating the reissued rule from further judicial review.[1]

Germane to this volume, press reaction to the high profile delisting suggests that it represents the introduction of politics into a listing process that until now has been based on science. For example the New York Times (2011) editorialized after the delisting that the ESA's "purpose is to base protections on science. Now that politics has been allowed to trump science when it comes to the gray wolf, which species will be next?" However, it is difficult to maintain that the congressional delisting of most Rocky Mountain gray wolves is the first time that politics has influenced which species are protected under the Act. For decades proponents and opponents of ESA listings have argued that politics influence federal agency determinations of which species are listed under the Act (see e.g., Houck 1993, 295; Doremus 1997, 1032, 1052; Biber & Brosi 2010, 372). Indeed the FWS rule reissued at the direction of Congress itself was regarded as at least partially politically motivated to appease western ranchers, hunters and state officials (DOI FWS 4/2009a; DOI FWS 5/2011b; Zuckerman 2010).

This chapter addresses the relationship between science and policy in the listing of species under the Endangered Species Act.[2] Under the Act species are to be listed by the Fish and Wildlife Service (FWS) or the National Oceanic and Atmospheric Administration (NOAA) as endangered or threatened "solely on the basis of the best scientific and commercial data available" (16 U.S.C. §1533[b][1][A]). The "solely" language indicates that science is to be the exclusive basis for listing decisions; "solely" was added in 1982 in the early years of the Reagan Administration specifically to prevent economic analysis from influencing listing determinations (Doremus 1997, 1054–1055; see also 50 C.F.R. § 424.11[c]). The permission to use the best commercial data available does not detract from the exclusive focus on science because the term merely allows the listing agencies to consider objective data about trade in species in evaluating threats they face and population trends (Doremus 1997, 1043). The ESA envisions that delisting decisions, like listing decisions, will be based exclusively on the best available scientific and commercial data (16 U.S.C. §1533[c][2]; 50 C.F.R. § 424.11[c] & [d]; see also Goble, 2009, 15–16). Nonetheless, it is well-recognized that decisions to list and delist species cannot, and are not, determined solely by science without input from policy considerations (see e.g., Doremus 1997; Doremus 2004, 402–3; Rohlf 2004; Carolan 2008b; Goble 2009; and the discussion below).

The complex interplay between science and policy in the ESA listing process has given rise to two concerns that mirror each other. The more familiar of these concerns is that the Act promotes a "science charade," to use a phrase Professor Wendy Wagner originally applied to the regulation of toxic risks (Wagner 1995), or the "scientizing" of politics in Professor Holly Doremus's words (Doremus 2005). The concern here is that the FWS and NOAA make listing (and delisting) decisions purporting to rely on science but in reality influenced strongly by unacknowledged policy considerations, with the attendant costs to democratic accountability and scientific progress. The other concern, perhaps less familiar in legal circles, is that the interplay the Act sets up politicizes science by inviting scientists to become advocates for or against species protection, and taking them away from their core expertise (see e.g., Ramey in this volume).

The dominant response to the twin concerns that the ESA listing process scientizes politics and politicizes science has been to argue that the scientific and policy considerations in listing decisions should be separated. "Separatist" proposals for disentangling science and policy in listing decisions come in two general forms. Some separatists argue for more transparent policy-oriented discussions about listing, an approach that might be termed "policy-zing" listing (see e.g., Doremus 1997). Others argue for continuing to think of listing primarily as a scientific exercise, but to distill out the politics embedded in ESA science to provide an improved, objective scientific basis for listing decisions, an approach that might be characterized as "depolicy-zing" listing (see e.g., Ramey in this volume). Proponents of improving the science used in ESA listing decisions are often regarded as using arguments for better science as a cover for reducing the protections that the ESA provides imperiled species. But neither approach is inherently more protective of species than the other; the ultimate impact would depend on the details of the proposal.

Toward the end of this chapter I argue that efforts to separate the science and policy aspects of listing decisions are likely to be in vain, and therefore advocate leaving the listing process alone. I instead propose reducing the momentousness of the listing decision as part of a broad package of policy changes that would have the beneficial by-product of lowering the significance of the scientization of politics and politicization of science associated with ESA listings. Under the existing ESA, the decision to list a species has high stakes because listing is the gateway to the potentially significant federal protections that the Act provides. The opponents of listing fear listing will impose a series of costly and rigid requirements on them from which they will be unable to escape. The proponents of listing fear that if listing is denied, species will lack adequate protection and continue to decline. While improving the protections species enjoy, we potentially could reduce the momentousness of the listing decisions in two ways: altering the consequences that flow from an ESA listing by amending the statute, and supplementing the protections that the ESA offers with other ambitious federal measures to protect biodiversity. Lowering the stakes associated with listing decisions in either or both of these ways could benefit species as well as lessen the significance of the scientization of politics and politicization of science that are likely hard-wired into ESA listing decisions.

THE INTERPLAY BETWEEN SCIENCE AND POLICY IN THE LISTING PROCESS

To understand the interplay of science and policy in ESA listing decisions, it is useful to review the process that the Act sets out for listing and delisting species, and the areas where policy necessarily comes into play because science alone cannot provide the answers.

The ESA assigns the duty to list "endangered" and "threatened" species to the Secretary of the Interior (16 U.S.C. § 1533[c]), while contemplating that the Secretary of the Interior and the Secretary of Commerce will share responsibility for evaluating whether species qualify (16 U.S.C. § 1533[a]). In practice, the Secretary of the Interior has delegated his responsibilities for implementing the ESA to the FWS, which is responsible for the vast share of the 1,374 U.S. listed species (DOI FWS 2011a). The Secretary of Commerce has delegated his ESA responsibilities to NOAA Fisheries, which is responsible for most anadromous and marine species. Although relatively few in number, these species include species that are controversial and costly to recover, such as Pacific salmon populations (see e.g. NOAA Fisheries; Marris 2007, 252). Species may come to the attention of FWS or NOAA as warranting listing either through the agencies'own initiative or through petitions which may come from any person (16 U.S.C. § 1533).

Under the Act, listing a species requires two major determinations (Doremus 1997, 1088). One is a taxonomic decision that the population of organisms at issue qualifies as a species under the Act. The other is a risk assessment that the population is sufficiently imperiled to qualify for protection under the Act. There is an argument that, textually, the Act requires only that the risk assessment, not the taxonomic, decision be made solely on the basis of the best available science (Doremus 1997, 1096–1096, 1138).[3] However, this interpretation does not seem to have been picked up in case law, and indeed there is case law applying the best available science standard to the taxonomic question (see e.g., *Center for Biological Diversity v. Lohn*, 296 F.Supp.2d 1223, 1236-1240 [W.D. Wash. 2003]).[4] Applying the best available science standard to the taxonomic as well as the risk assessment determinations is likely an accurate interpretation of Congressional intent as reflected in the legislative history of the Act, which seems to envision that listing determinations will be a scientific exercise (Doremus 1997, 1096).

Neither the taxonomic decision nor the risk assessment underlying a listing decision is a purely scientific exercise, however. Starting with the taxonomic question, science often will be unable to determine whether a population of organisms is a species as defined in the Act. The Act defines a species as including "any subspecies of fish or wildlife or plants, and any distinct population segment of any species of vertebrate fish or wildlife which interbreeds when mature" (16 U.S.C. § 1532[16]). Thus species under the ESA include species, subspecies, and distinct population segments (DPSs) of vertebrate fish and wildlife. However, there is no single scientific definition of any of these three terms.

There are over twenty concepts of species in modern scientific literature (Mayden 1997, 389). Different definitions may have radically different implications for whether a population is recognized as a species. The biological species concept (BSC) is "probably the most widely used species definition" (Doremus 2010, 184). The second most widely used species concept is the phylogenetic species concept (PSC), which by one account generates "48 percent more species than the biological species concept does for the same group of organisms" (Marris 2007, 251 [citing Agapow et al. 2004]). Due at least partly to the application of the PSC, many subspecies have been elevated to species status recently in a phenomenon dubbed "taxonomic inflation" (Isaac 2004; Marris 2007; Ramey 2010, 4–9).

Many of the controversies about whether groups are protectable entities under the ESA have concerned listing groups below the species level (GAO 2003, 58-60; Rohlf 2004, 488–90, 511–20). Science provides even less guidance on what counts as a subspecies than a species because there is little agreement on the degree of differentiation required for a population to count as a separate subspecies (see e.g., Doremus 2010, Doremus 1997, Cronin 2007, Cronin 1997). Moreover, the FWS and NOAA have not articulated clear policies for delineating subspecies (Doremus 2010, 192). Distinct population segment is not even a scientific term; it had not been used in scientific literature before it was incorporated into the ESA in 1978 (Brosnan 2007, 994). In a joint policy, FWS and NOAA have articulated requirements for recognizing a population as a DPS (DOI FWS & DOC NOAA 1996). The DPS policy "is consistent with" an earlier NOAA policy defining "evolutionary significant units" (ESUs) (DOI FWS & DOC NOAA 1996, 4722; DOC NOAA 1991).[5] Robin Waples, the person who developed the underpinnings of the ESU policy, frankly admits the role of policy considerations in diagnosing DPSs.[6]

Turning to the decision that a species is sufficiently imperiled to warrant listing, the ESA guides the risk assessment decision to a degree. It specifies the kind of risk a species must face to be listed (a danger of extinction for endangered species, and the likelihood of becoming in danger of extinction for threatened species), over what time period the species must face that risk (immediately in the case of endangered species, and "within the foreseeable future" for threatened species), and where the species should face that risk ("throughout all or a significant portion of its range") (16 U.S.C. § 1532[6] and [20]; see also Goble 2009, 7).[7] In a provision that may obscure the need to make a risk assessment to list a species, the Act requires that the listing agency identify whether a species is endangered or threatened because of any of five threats, any one of which is sufficient for the species to be listed (16 U.S.C. 1533[a][1], excerpted in note 3; Goble 2009, 8, 14). The listing agency also is required to take into account "those efforts, if any, being made by any State or foreign nation, or any political subdivision of a State or foreign nation, to protect [the] . . . species" (16 U.S.C. 1533[b][1][A]).

Science may be able to provide information relevant to the risk assessment decision underlying a listing determination. For example, scientists may be able to estimate a species' extinction risk over a specified period of time. Assuming they are available, however, estimates of extinction risk are likely to be subject to uncertainties and to incorporate assumptions that themselves may reflect policy choices (Doremus 1997, 1118, 1121). Moreover, estimates of extinction risk cannot by themselves determine whether a species should be listed. Determining that a species should be listed requires a policy decision that the extinction risk a species faces—whether it is estimable or not—is unacceptable from a societal point of view (see also Doremus 1997; Carolan 2008a; Goble 2009; Rohlf 2004, 502 n. 73 and accompanying text, 505–507). As Carolan (2008a, 947) aptly explains, "deciding when a species is safe versus endangered (and thus in need of protection) is really a question of how much risk a society is willing to take with that species. And since there is no 'correct' level of risk, such decisions rest upon policy rather than scientific choices."

A species may be delisted if it is no longer endangered or threatened for any of three reasons: its population is recovered to the point that it no longer requires the Act's protections, the species is extinct, or the species was listed in error (50 C.F.R. § 424.11[d]). As previously indicated, the ESA directs that like listing decisions, delistings must be made solely on the basis of the best scientific and commercial data available (16 U.S.C. §1533[c][2]; 50 C.F.R. § 424.11[c] and [d]; see also Goble 2009, 15–16). There also is significant scope for policy considerations to come into play in delisting decisions.

Central to the decision to delist on the basis that a species is recovered is a risk assessment that focuses on whether the species' population has progressed to the point that it can survive without the ESA's protections. In undertaking this risk assessment, the listing agencies are required to consider the status of the species in light of the same five factors used in deciding to list the species (16 U.S.C. 1533[a][1], excerpted in note 3; 16 U.S.C. §1533[c][2]; 50 C.F.R. § 424.11[c] and [d]; Goble 2009, 15–16). In practice, the risk assessment at the delisting stage seems to devolve into a "demographic" inquiry into the status and dispersal of the population, and a "risk management inquiry" into the protections that will exist for the species in the absence of the ESA (Goble 2009, 17–18 [indicating that "recovery has two elements"]; id. at 40–44). But determining whether to take the risk of delisting a species is at heart a value judgment about the degree of precaution that should be taken to safeguard a species, much like the risk assessment that must be made in listing the species in the first place (Goble 2009, 18).

At first glance, a taxonomic inquiry would seem unlikely at the delisting stage because the initial decision to list a population entails a decision that it counts as a species under the ESA. But there might be a taxonomic decision as part of a decision to delist a species, again with the scope for policy considerations to come into play. For example, in delisting the agency might redefine the species, perhaps carving the species into distinct DPSs and then treating the DPSs distinctly, as the FWS has done with the gray wolf.[8] Alternatively, there might be a taxonomic decision as part of a delisting because the FWS or NOAA is contemplating delisting the species on the basis that it was listed in error because it is not a species. Below I refer to the debate in the 2000s about whether the Preble's meadow jumping mouse is a subspecies which occurred in the context of the FWS deciding whether to delist the mouse on the basis that the original listing reflected a taxonomic error (DOI FWS 2/2005).

TWIN CONCERNS ABOUT THE INTERPLAY OF SCIENCE AND POLICY

Thus the listing and delisting processes that the ESA establishes, although intended by Congress to be based on the best available science, include considerable scope for policy considerations. As mentioned earlier, the interplay of science and policy in listing decisions has given rise to twin concerns about the one infecting the other: a concern that FWS and NOAA are "scientizizing" listing decisions that ultimately rest on policy judgments, and a concern that science is being "politicized" by its involvement in listing decisions. I now turn to these concerns and then discuss the dominant responses to them.

The better documented of the two concerns in legal scholarship is the concern about the scientizizing of politics. In a well-known 1995 article, Professor Wagner argued that a "science charade" was impeding the regulation of toxics (Wagner 1995, 1616–1617). Wagner defined the "science charade" as "where agencies exaggerate the contributions made by science in setting toxic standards in order to avoid accountability for the underlying policy decisions" (Wagner 1995, 1617). Wagner was careful to suggest that regulatory agencies engaged in the charade were not always consciously trying to pull the wool over anybody's eyes; she argued that regulators might engage in the charade unintentionally, intentionally or in a premeditated manner (Wager 1995, 1628–1650). Two years later, in what remains the leading law review article on the ESA listing process, Professor Doremus invoked Wagner's term and argued that "the ESA's current requirement that listing decisions rest solely on science . . . has forced the listing agencies into a 'science charade,' in which they must pretend to make non-scientific decisions entirely on the basis of science" (Doremus 1997, 1035). Doremus highlighted the policy judgments underlying the two components of a listing decision described above, the taxonomic decision that a species should be protected under the Act, and what she called the viability determination (and I referred to as a risk assessment) that the risk of the species becoming extinct is too great (Doremus 1997, 1034–1035).

More recently, Goble (2009) has emphasized that the decision to delist a species on the basis that it is recovered is an "ethical/policy judgment" (18) about who should bear the risk that an error is made in assessing whether the species would go extinct once the ESA's protections are removed: "species or human activities" (14). Goble's description of delisting is picked up in the 2010 federal district court decision, targeted by the 2011 appropriations rider, thwarting the FWS's effort to delist the northern Rocky Mountain wolf. In words that would please the legal realist, Judge Donald W. Molloy states that "it is not necessarily the case that threatened or endangered status can be determined solely on the basis of scientific evidence alone. Beyond the question of risk is the issue of the acceptability of risk. . . . The decision that a risk is acceptable regarding a specific species is, in turn, an ethical or policy judgment" (*Defenders of Wildlife v. Salazar*, 729 F.Supp.2d 1207, 1210 [D.Mont. 2010] [citing Goble 2010]).

As if it were necessary, the delisting of the northern Rocky Mountain gray wolf distinct population segment provides fresh evidence of the potential for invocations of science to obscure policy judgments or, to use Doremus's terms, for policy to become scientized. Consider the definition of the recovery goal in the FWS's 2009 rule delisting the northern Rocky Mountain gray wolf, the rule that was blocked by Judge Molloy's 2010 decision and then implemented pursuant to the 2011 appropriations measure (DOI FWS 4/2009a).

The FWS rule argued that the wolf population had recovered to the point that it no longer requires the Act's protections. In doing so, the FWS maintained the definition of what would count as a recovered northern Rocky Mountain gray wolf DPS that the agency had first adopted in 1994, before wolves were reintroduced to the northern Rocky Mountains by bringing in wolves from Canada in 1995 and 1996. Under this definition, a recovered Rocky Mountain gray wolf population is '"[t]hirty or more breeding pairs comprising some 300+ wolves in a metapopulation (a population that exists as partially isolated sets of subpopulations) with genetic exchange between subpopulations'" (DOI FWS 4/2009a, 15134 [quoting Service 1994 p. 6:75]).[9] As of 2008, the FWS estimated the number of breeding pairs was approximately three times the number required by the definition (95 vs. 30) and five times greater than the overall population goal (1,639 vs. 300) (15123).[10] To provide a margin of safety, the FWS's post-delisting population goals also exceed the recovery goal: each of Montana, Idaho, and Wyoming are to have "at least 15 breeding pairs and at least 150 wolves in mid-winter" (15132). Post-delisting, the FWS, the National Park Service, and the states will manage for a population averaging over 1,100 wolves (15137).

Going back to 1994, the FWS was aware that there is an argument grounded in conservation biology that a population of adult breeding animals numbering 500 is necessary for the long-term viability of the Rocky Mountain gray wolf (see DOI FWS 1994, Appendix 9, 38). This population of breeders would yield a total population of perhaps 2,500 to 5,000 wolves (Harbine 2009, 199; see also DOI FWS 1994, Appendix 9, 38 [an N_e of 500 "would equate to a total population in the low thousands"]).[11] The FWS also received comments prior to publishing the 2009 delisting rule arguing for a higher recovery goal—some suggested an overall population of "2,000 to 6,000 or more" wolves is necessary for a viable population (DOI FWS 4/2009a, 15140).

Why did the FWS maintain the recovery goal of 300 rather than adopt a higher goal? The 2009 delisting rule initially seems to imply that it was science that drove the decision to maintain the recovery goal. Invoking the authority of science, the rule indicates that prior to adopting the definition in 1994, "the Service conducted a thorough literature review of wolf population viability analysis and minimum viable populations, reviewed the recovery goals for other wolf populations, surveyed the opinions of the top 43 wolf experts in North America, of which 25 responded, and incorporated [the agency's] . . . own expertise" (DOI FWS 4/2009a, 15130). The rule also indicates that in 2001–2002 the FWS reviewed its 1994 recovery goal. Again, the review is described as a scientific exercise, with the rule mentioning that it involved another literature review and attempts to re-survey the same forty-three experts the FWS contacted in 1994 plus attempts to survey an additional forty-three other biologists in North America and Europe, of who fifty-three responded (15131). According to the rule, 78 percent of the experts who responded as part of the 2001-2002 review agreed that the FWS's goal "represented a viable wolf population," albeit with qualifications (15131). Finally, the rule indicates that the FWS reviewed its recovery goal once again before issuing the 2009 rule and that it "continue[s] to conclude, as did over three-fourths of the experts contacted, that the recovery goal is adequate to ensure wolves in the NRM [Northern Rocky Mountains] do not again become threatened or endangered" (15140).[12]

Notwithstanding the effort to portray the recovery goal as a scientific choice, however, there are reasons for thinking that the FWS likely kept the 1994 recovery goal because the agency judged that the political and economic costs of a higher target would be too great. The 2009 rule argues that the current wolf population of 1,639 in 2008 approaches the carrying capacity of existing suitable habitat (15138). It suggests that existing suitable habitat would not be able to support the higher overall population target of 2,000 to 6,000 suggested by some commenters "without tolerating rates of livestock depredation and impacts to big game populations many times higher than has occurred in the past twenty years" (15140). Furthermore the FWS alludes to the political and economic costs of creating additional suitable habitat, warning that "would require massive Federal and State programs to reduce or eliminate livestock on Federal, State, Tribal and, mostly, private property" (15140). Against the backdrop of these statements, the decision to stick with a recovery goal of 300 seems to represent a decision informed by science, but resting ultimately on a policy calculus about the extent to which the FWS could push the protection of wolves in a part of the country where there is considerable hostility to wolves, especially from ranchers and hunters (see also Harbine 2009, 201, who argues that the recovery goal was politically established; Bergstrom et al. 2009). However, as befits a rule produced by an agency that is "scientizing politics," the 2009 delisting rule is less than explicit in acknowledging the role of policy considerations in its determinations.[13]

While the interplay between science and policy in ESA listing decisions gives rise to concerns that science is used by FWS and NOAA to shield policy conclusions from scrutiny, the interplay also generates concerns that scientists doing ESA-related work have become politicized. In Roger Pielke Jr.'s (2007) work on the relationship between science and policy, he argues that one way that scientists can relate to the policy process is as "Issue Advocates." According to Pielke Jr. (2007, 15), "[t]he Issue Advocate aligns him/herself with a group (or faction) seeking to advance its interests through policy and politics. The Issue Advocate accepts the notion that science must be engaged with decision-makers and seeks to participate in the decision-making process."[14] Scientists might not be aware they are acting as issue advocates. Many conservation biologists are probably favorably inclined to protect populations under the ESA because many of them entered biology because of their affection for nature and their disciplinary training likely reinforces their predisposition to protect biodiversity.[15] Many biologists are probably unaware of the value judgments embedded in their work (Doremus 2005, 261).

In this volume Dr. Rob Ramey voices the concern that the ESA is politicizing science, although he does not use this phrase. Ramey suggests that biologists sometimes act as advocates on behalf of listing species without clearly acknowledging their advocacy—in other words, as "Stealth Issue Advocates" in Pielke Jr.'s typology.[16] For example Ramey refers to "individuals" forming "'species cartels'" that "can influence the peer review process, provide a united front to advocate for particular decisions, and repress the publication of information that does not agree with their positions" (Ramey 2010, 12). Ramey himself was accused of "advocacy masquerading as science" for reduced species protection during the debate about whether the Preble's meadow jumping mouse should be delisted on the basis that it is not a distinct subspecies, in which Ramey's research concluding that the Preble's is not a valid subspecies played a leading role (Ramey et al. 2005; Martin 2006, 248).

Indeed the controversy in the 2000s about whether to delist the Preble's on the basis that it is not a subspecies vividly illustrates a number of ways in which policy considerations are present in ESA listing related science. The mouse was listed as threatened subspecies in 1998, affecting development in Colorado and Wyoming, especially after over 31,000 acres in the two states were designated as critical habitat for the Preble's (Marris 2007, 250; Crifasi 2007, 517-518; Carolan 2008a, 947). In

2002, Ramey proposed a study to reexamine whether the Preble's is a subspecies (Holthouse 2005). The State of Wyoming and FWS agreed to fund it, something they probably would not have done if the Preble's had not been listed under the Act already (Crifasi 2007, 519). Wyoming in particular likely funded the study because the state had been seeking to delist the species for years (Holthouse 2005). In 2005, in response to petitions from the group Coloradans for Water Conservation and Development and the State of Wyoming's Office of the Governor, the FWS proposed to delist the subspecies, arguing substantially based on Ramey's work that the mouse is not a subspecies (DOI FWS, 2/2005). After controversy erupted, the FWS commissioned additional research from another team of biologists who disagreed with Ramey's conclusion that the mouse was not a subspecies (King et al. 2006; Crifasi 2007, 521). The agency then commissioned another scientific review to evaluate the work of both sides on the taxonomic status of the mouse (Arbogast et al. 2006). The panel sided with the group that had concluded that the mouse still should be considered a subspecies.[17] Thus in a tangible sense the ESA itself spawned two original research projects into the status of the Preble's (Ramey et al. 2005; King et al. 2006) as well as much peer review (Crifasi 2007, 520-521), providing ample evidence for Ramey's point in this volume that the Act has "catapulted taxonomy from" obscurity (Ramey 2010, 3).[18]

Substantively, the debate in the 2000s about whether the Preble's is a valid subspecies underscores the potential for scientists working on ESA listing related research to become enmeshed in debates that ultimately turn on values, not on scientific findings. In explaining why Ramey et al. and King et al. came to different conclusions about whether the Preble's is a distinct subspecies, the scientific panel that the FWS hired to review both groups' research concluded that "[t]here are multiple philosophical differences between KEA [King et al. 2006] and REA [Ramey et al. 2005] that contribute to their difference of opinion regarding the taxonomy of Z. h. preblei" (Arbogast et al. 2006, 35).[19] Philosophical differences in this context are what this chapter refers to as policy differences.

Among the policy differences that the panel identified between KEA and REA are different definitions of subspecies (Arbogast et al. 2006, 35). As discussed above, there is no scientific consensus on the degree of differentiation required between groups to label them distinct subspecies. The panel observed that "neither REA nor KEA rigorously employed a definition of 'subspecies' per se from the literature" but rather "both use a definition of ESU as a proxy for subspecies, even citing some of the same references in doing so" (38). Where the authors differ is in the way they translate the "definitions of ESUs into a testable definition of subspecies" (38). REA adopted a more stringent definition of a subspecies than KEA (38),[20] a point acknowledged by REA (Ramey et al. 2007, 1) and KEA (King et al. 2006, 4347-4348).

Another philosophical difference the panel identified is the "choice of null or alternative hypotheses and the 'burden of proof'" (Arbogast et al. 2006, 35). For REA, the null hypothesis is that the Preble's is not a distinct subspecies, and the burden of proof rests on showing evidence that the Preble's is a distinct subspecies (Ramey et al. 2005, 330; see also Carolan 2008b, 460). For KEA on the other hand, the null hypothesis is that the Preble's is a distinct subspecies and the burden of proof rests on showing that it is not a distinct subspecies (see e.g., King et al. 2006, 4332). The panel sided with KEA on where the burden of proof should lie, reasoning that "[b]ecause Z. h. preblei is a formally described, valid, and commonly recognized taxon, we concluded that the burden of proof should lie in clearly showing that its taxonomic status is not warranted" (Arbogast et al. 2006, 39).[21]

What is largely unacknowledged in the scientific literature on the status of the Preble's is the "differing visions of conservation policy" that underlie the differences between REA and KEA (and the panel to the extent that it sides with KEA) on issues such as the testable definition of species, the null hypothesis and the burden of proof (Carolan 2008a, 948).[22] Ramey et al. seem to adopt a more stringent definition of subspecies and a null hypothesis that requires proving that the Preble's is a distinct subspecies based on a value judgment that conservation policy should aim to protect species that are clearly genetically distinct from others, or as Ramey put it colloquially to a journalist, "the big branches" rather than "the little twigs" if you think of "biodiversity as a tree that's in trouble" (Holthouse 2005; see also Ramey et al. 2005, 329; Ramey et al. 2006, 246).[23] This is a highly defensible position. As Ramey explains, there is a limited amount of money for conservation. It may make sense to list and direct ESA dollars to protecting the populations that are more clearly genetically distinct from others, if we assume as Ramey seems to do that the goal of the ESA is protecting genetic diversity. But it is important to recognize that the idea that the ESA should protect genetic diversity is a value judgment not shared by everyone, and not clearly reflected in the ESA, which refers to the "esthetic, ecological, educational, historical, recreational, and scientific value" of species (see ESA § 2[a][3]; Doremus 1997, 1107, 1132; Doremus 2010, 217–23).

King et al.'s looser definition of subspecies and different choice of null hypothesis and burden of proof reflect a different value judgment that the ESA should protect a broad array of populations, even if they are not clearly genetically distinct at present, as a precaution to allow as much room as possible for evolution (King et al. 2006, 4347–4348). This is also a highly defensible position, and it may be more consistent than Ramey's with Congressional attitudes toward the ESA, at least in the 1970s and 1980s. But due to the scarcity of resources available to protect biodiversity it is probably not possible for the ESA to protect the full array of populations that this approach would encompass. Moreover King et al.'s approach inhibits using the listing process to set priorities for allocating the limited available funding for species protection (Ramey et al. 2006, 246). Again, though, it is important to acknowledge that King et al.'s position, like that of Ramey et al., is a value judgment that needs to be argued for, not a scientific claim that is falsifiable. Thus the controversy about whether the Preble's is a distinct subspecies is an example of scientists debating conservation policy objectives through the lens of science—the politicization of science.

DOMINANT RESPONSES AND THEIR LIMITS

The scientization of politics and the politicization of science evident in the wolf and Preble's cases are problematic partly because of their implications for the listing process.[24] They may inhibit public participation in, and public accountability for, listing decisions. When regulators and scientists debate policy issues in scientific language that is not easily accessible, they make it harder for the lay public to participate meaningfully in debates and hold decision-makers accountable for choices that ultimately turn on values, not complex science (Wagner 1995, Carolan 2008a). Relatedly, justifying listing decisions as resting on science may dissuade the judiciary from closely scrutinizing listing decisions, although the evidence is mixed about the degree of deference courts give agencies for decisions they claim are based on science (see Meazell 2010-2011, 738, 776; Doremus 2004, 429–32). In addition, thinking about issues that are fundamentally values issues as scientific issues may lead governments and others to invest unnecessarily in science that will not help to resolve the issues (for a similar argument referring to the regulation of toxics, see Wagner 1995, 1677–1685). As mentioned above, the Preble's controversy spawned a small industry of scientific studies and peer review.

The scientization of politics and politicization of science also may be detrimental to science. Regulators and scientists may set to the side gaps and uncertainties that should be the subject of additional scientific research in a rush to suggest that science clearly justifies their conclusions (Wagner 1995, 1687). Debating whether species should be listed as if the decision is primarily a scientific choice also may undermine public esteem for scientists. If the debate degenerates into a battle of dueling scientists, members of the public may lose respect for scientists generally and come to regard them as paid experts who will take the side of their paymasters (Wagner 1995, 1688; Crifasi 2007, 528–29; Brosnan 2007, 993 note 22 [referring to "dueling scientists" in the debate about whether the Preble's is a distinct subspecies]).

Most of the responses to the twin phenomena of the politicization of science and the scientization of politics have argued for changes to the listing process that would separate the scientific and policy considerations in the process. These separatist responses have taken two general forms.

One set of proposals would more overtly "policy-cize" the listing process by dropping the pretense that listing decisions should be based solely on the best available science. For example, in the legal literature Professor Doremus (1997, 1153) argues that Congress "should separate the scientific aspects of listing determinations from the value judgments" (1135). Doremus views the taxonomic question of which groups to protect as an issue that should not be resolved by science alone. Under her proposal, groups that taxonomists consider full species would be protected (1135). Groups below the species level would be eligible for listing if they were "identifiable" and "significantly different from other groups with respect to esthetic, ecological, educational, historical, recreational, or scientific value" (1136). Doremus would not limit the agencies to considering only science in determining which groups to protect but she would prohibit the agencies from taking into account the economic costs of listing (1137–1138). Doremus regards the risk assessment question as potentially determinable by science if Congress (or the agencies) more clearly specify the extinction risk that species must face and the relevant time period over which they must face for it (1142–1144). With clearer specifications of these issues in hand, Doremus envisions the agencies relying solely on scientific information to determine if species qualify as endangered or threatened (1142–1145).[25]

A second set of responses retain the idea that listing decisions should be made based exclusively on science and focuses on "depoliticizing" the science relevant to ESA listing decisions. Ramey's chapter in this volume falls into this category.[26] Ramey suggests that the taxonomic question of which groups should be protected under the Act still should be decided by science; unlike Doremus, he does not envisage groups being protected after a policy debate that would allow consideration of groups' "esthetic, ecological, educational, historical [or recreational] . . . value." However, he argues that the science used to determine whether groups should be protected needs to be cleansed of its current "subjectivity" (Ramey 2010, 21) by developing clear thresholds of the degree of differentiation that a group must have from other groups to be listed.[27] Ramey acknowledges that developing these thresholds would involve debate (18). In this chapter's terms it would involve policy decisions, because it is not self-evident what degree of distinctness is required from other groups to be a protected group under the ESA. But once the thresholds were developed, then science could determine which groups should be listed objectively by applying the thresholds to the data (18).

While vulnerable to distinct criticisms,[28] separatist proposals such as Doremus's and Ramey's share a common weakness: they presume that it is possible to sufficiently separate science from policy to confine the two to their own realms. However efforts to disentangle the two to any significant degree are likely to be in vain. As Brosnan (2007, 998) argues, "[a]lthough we may be keen to distinguish between science and policy, the lines between them are becoming less and less distinct."

To take the science side of the equation first, value judgments are probably so deeply entrenched in environmental science that it would be extremely difficult to cleanse it of policy considerations (see Biber 2011, 13–18). Assume for the sake of argument that Congress would be willing to clearly define thresholds of differentiation as Ramey recommends, or a required level of extinction as Doremus proposes. Scientists are nonetheless likely to encounter many gaps in data and uncertainties in applying these definitions because uncertainty is pervasive in ESA science. For example, relevant to any given listing decision, there may be uncertainty about the present and likely future population of a species, the life cycle of the species or the threats facing the species.[29] The choices scientists make in dealing with these gaps and uncertainties "are not 'scientific' as the public understands that term, although they are familiar to scientists and indeed are an unavoidable part of the scientific enterprise" (Doremus 2004, 438). Instead these kinds of choices are value choices that reflect views about the degree of certainty required as a predicate for regulatory action that may save species but also impose significant costs on private and public actors (Doremus 2004, 439).

Assume for the moment that it would be possible to bring out the policy choices embedded in ESA science or at least the most important of them. To distill the policy from the science also would require awareness among scientists of the value choices embedded in their work, something which as mentioned above cannot be assumed. In addition, there would need to be an incentive for scientists to acknowledge openly the role of policy in ESA related science. But acknowledging this role carries the risk for science of undermining its authority. It is far from clear that scientists have any particular claim to expertise in making policy decisions about issues such as the degree of certainty that should be required to act to protect species.

In addition to the problems with removing policy judgments from ESA related science, it will be difficult to take the science out of the policy-making process. The best available science has been the exclusive legal basis for listing species for almost three decades, and science now is deeply embedded in FWS's and NOAA's listing determinations, though it is not necessarily determinative of these decisions for the reasons already discussed. Recall FWS's extensive invocation of science in justifying the use of the 1994 recovery goal in the 2009 Rocky Mountain wolf delisting rule.

It is not only inertia that may make it hard to separate the scientific and policy aspects of listing decisions. The listing agencies also have powerful incentives to continue to invoke the authority of science to justify policy decisions. Invoking science may enable them to dodge responsibility for listing decisions because, as mentioned above, it is often difficult for lay members of the public to acquire sufficient understanding of the science to be able to grasp what it does and does not determine (see Wagner 1995, Doremus 1997; but see Doremus 2004, 426–29). Perhaps more importantly science serves a fundamental bureaucratic need of agencies by providing them with a framework that they can use to structure their decision-making. Without science as a backbone for their decision-making, the FWS and NOAA might be left rudderless in a sea of open-ended policy considerations in deciding whether to list a species.

Congress also may have powerful incentives to perpetuate the science charade at the agency level (Wagner 1995, 1704–1705). In theory Congress could end the charade at the agency level by deciding to list species itself. But this would make Congress directly accountable for listing decisions, something that Congress generally does not seem enthusiastic about being. Indeed the recent Congressional delisting of most Rocky Mountain gray wolves apparently is the first successful "legislative effort regarding listing of any species" (CRS 2011, 8). Direct congressional listing also would be impractical on a routine scale given the many imperiled species and Congress's workload (Wagner 1995, 1704–1705). Without taking listing decisions upon itself, Congress could end the science charade at the agency level by setting clearer standards for the FWS and NOAA to apply or by directing the agencies to consider policy considerations in listing and not confining the agencies exclusively to the best available science. But there are many advantages to Congress of the status quo under which it delegates listing decisions to the agencies and directs the agencies to act based exclusively on the best available science. It enables Congress to avoid the politically contentious task of setting clearer standards. The direction to base listing decisions on science requires the agencies to use a body of knowledge that many members of the public continue to regard as providing an objective, ideology-free basis for decision making.[30]

In thinking about the incentives of Congress and the listing agencies, it is also worth considering that neither the agencies nor Congress face much political pressure to stop the science charade at the agency level. Environmentalists and the regulated community seem quite content with the status quo under which science is supposed to be the basis for listing decisions but policy enters the picture. Environmentalists continue to regard science as a body of knowledge that favors conservation, reflecting perhaps the conservationist leanings of many scientists and the values built into disciplines such as conservation biology (see generally Doremus 2005, 258–61). On the other side, regulated actors have become adept at using science to advance their interests, for example by emphasizing the gaps and uncertainties in science and the inadequacies of methodologies and urging greater use of peer review (see generally Doremus 2004, 415; Doremus 2005, 261–66). Outside of academic and policy circles, it is hard to identify a powerful political constituency for ending the scientization of politics.

Other Options

Given the obstacles to disentangling the policy and scientific aspects of listing and confining them to separate domains, I suggest that as a second-best we attempt to reduce the significance of the scientization of politics and politicization of science as a by-product of policy changes to improve the outcomes for species. It is well known that the ESA has had a mixed record in helping species.[31] Many imperiled species currently are not listed, with the lists accounting for perhaps only 15 to 20 percent of imperiled species (Davis et al. 2006, 297). Of the species that have been listed, very few species have been delisted because their populations have recovered to the point that they can be viable without the Act's protections (see e.g., DOI FWS 2011b; Scott et al. 2006, 31–33). The Act, however, probably has prevented many listed species from going extinct; listed populations generally stabilize or increase the longer the time a species is listed (Scott et al. 2006, 31–32; Male & Bates 2005, 988; Gifford 2007, 4). As a consequence of broader efforts to improve outcomes for species, we might reduce the significance of the politicization and scientization phenomena by lowering the importance of the decision to list a species under the ESA.

Under the existing ESA, the decision whether to list or not list a species is a momentous decision because listing is the gateway to the Act's potentially significant protections. Once a species is listed, a standard series of statutory protections kick in to protect that species. The listing agency is charged with designating any critical habit for the species concurrently with listing unless an exception applies (16 U.S.C. §1533[a][3][A] and [b][6][C]). Federal agencies must "insure that any action authorized, funded, or carried out by such agency . . . is not likely to jeopardize the continued existence of any endangered species or threatened species or result in the destruction or adverse modification of" the species' critical habitat (16 U.S.C. § 1536[a][2]). Persons generally are prohibited from taking any endangered species (16 U.S.C. § 1538[a][1][B]). The taking of threatened species also may be prohibited by regulation (16 U.S.C. §1533[d]).[32] The listing agency also must develop and implement recovery plans for listed species, although recovery plans do not create judicially enforceable obligations (16 U.S.C. §1533[f]; *Fund for Animals v. Rice*, 85 F.3d 535, 547–48 [11th Cir. 1996]; Quarles 2010, 10815 note 3).

Environmentalists often worry that if imperiled species are not listed the species and the ecosystems in which they exist will continue to decline without federal protection. Thus environmentalists often provide the impetus to list species, filing listing petitions and litigating when the services fail to act on these petitions or decline to list species (see e.g.,

Biber & Brosi 2010, 344 n.111). It is pressure from a number of environmental groups to list over 1,230 species that recently led the FWS to take desperate measures to bring its listing workload under control (see e.g., Woody 2011; Maron 2011).[33] Conversely, listing often is opposed by regulated actors who fear they will bear most of the costs of ESA protections. The Preble's controversy is illustrative. For environmentalists, the listing of the Preble's is a means of protecting areas subject to development pressure as well as the mouse (Crifasi 2007, 515, 528). For developers and other commercial interests on the other hand, the listing is an impediment to economic activity whose costs will disproportionately fall on them (Crifasi 2007, 516–18, 528). Indeed the FWS reportedly estimated that it cost real estate developers in Wyoming and Colorado $8 to $18 million annually in just consulting fees to protect the habitat of the Preble's (Holthouse 2005).

There are a number of ways that we might reduce the significance of the decision to list or not a list a species and by extension the significance of the politicization of ESA related science and scientization of ESA listing decisions—while also improving the protections available to species. One possibility would be to change the Act to delink the protections that a species enjoys from the listing decision, without eviscerating the Act's protections. Another possibility would be to expand the use of non-ESA frameworks to protect biodiversity and therefore to reduce the importance of being listed under the ESA. These options could be pursued alone or together.

Let's turn first to options for reforming the ESA that while aiming to improve protections for species might have the side effect of lowering the stakes of the listing decision. One option I have proposed elsewhere is to restructure the consequences of listing a species so that listing would no longer trigger the standard package of protections described above (Wyman 2008). Instead listing would trigger a judicially enforceable requirement that the listing agency develop and implement the most cost-effective mechanisms for recovering the species to the point that it no longer requires the Act's protections, a stated goal of the Act (16 U.S.C. §1531[b] and §1532[3]).[34] These mechanisms could include variations on the existing protections in the ESA, but they also should include more flexible and market-based measures developed over the past decades under the ESA, such as land purchases, conservation payments, conservation easements, conservation banking, recovery credit systems, recovery and habitat conservation plans, and fees for converting the habitat of endangered and threatened species. In identifying the most cost-effective mechanisms for protecting the species, the listing agency could take a broad view of what counts as a cost, and count not only easily monetizable costs but also ethical and political costs. The agency could be required to finalize the cost-effective protections within a legislated timeframe to avoid delays that complicate efforts to protect the species. While the listing agency determines the most cost-effective means of protecting the species, an interim package of protections, likely mandated by legislation for administrative convenience, could come into effect to prevent the species and its habitat from further deteriorating.

Decoupling the listing decision from the decision about what protections a species enjoys along the lines described above might enable the implementation of measures more closely tailored to individual species' needs than the current Act allows, and therefore improve the outcomes for species listed under the Act. It also would require the use of a cost-effectiveness test in determining permanent protections, which might promote a more rational allocation of the limited funds available for conserving species and therefore increase the overall level of species protection. In addition, the decoupling proposal might have the side effect of reducing the significance of the listing decision for species that are listed because the fact of listing would no longer trigger the standard series of protections in the current ESA, but rather an interim series of protections with the permanent protections to be developed later by the FWS. On the other hand, the listing decision still would be important for species that are listed because it is this decision that would trigger the interim protections and the obligation to develop and implement the cost-effective protection measures. Indeed, one critic of the decoupling proposal suggested that the uncertainty that would exist while the listing agency defined the ongoing protections would make the listing decision "more momentous, not less" (Quarles 2010, 10817). Moreover, for species that an agency decided not to list, the listing (or rather, no-listing) decision would remain equally significant because the species would enjoy no additional protections following that decision.

In criticizing my decoupling proposal, Quarles (2010, 10818) outlines another option for reforming the ESA that might have the effect of more significantly reducing the importance of the listing decision. As Quarles explains, the idea of giving legal effect to recovery plans has been raised in past discussions of reforming the ESA but not greeted with enthusiasm by the regulated community that fears "interpos[ing] the most stringent ESA standard—recovery—in the existing protective mechanisms" (10818).[35] But giving legal effect to recovery plans might be more palatable for regulated actors if combined with a more realistic objective for the plans than recovering listed species to the point that they no longer require the Act's protections. Quarles does not specify what that less ambitious objective might be, and there is a danger that if it is set too low that it would eviscerate the utility of legally enforceable recovery plans (for a related concern, see Irvin 2010, 10810). But given that recovery plans currently do not have the force of law, there would seem to be room to improve the protections available to species by making the plans legally enforceable with a new goal that falls somewhat, although not too far, short of recovery. Remember that the newly enforceable recovery plans would be in addition to the existing protective mechanisms that aim for something less than recovery (Quarles 2010, 10818). Quarles anticipates that measures in the enforceable plans would influence the requirements of the other protective mechanisms such as section 7 (Quarles 2010, 10818).

Quarles's proposal would reduce the momentousness of ESA listing for listed species by lowering the goal for listed species. How much less momentous the decision to list would become would depend on the ambition of the reformulated objective. However, even under Quarles's proposal, listing still would be meaningful because it would remain necessary to secure a recovery plan (the plans would likely have to be renamed to reflect the new objective) and the Act's existing protections, and there would be uncertainty about what protections would apply for the species while the "recovery" plan was under development. In addition, for species that the agencies decided not to list, the no-listing decision would retain its current significance because they would benefit from no additional federal protections above what they currently enjoy. Moreover any proposal to amend the ESA is likely to be politically difficult to implement, as evidenced by the fact that "the ESA has been substantively amended just once in the last 20 years" even though volumes have been published about its shortcomings (Johnson 2010, 10812).

Let's turn then to a policy option that might improve the protections for species without touching the ESA, and might have the side benefit of reducing the significance of ESA listing decisions and by extension the politicization and scientizization associated with them. One standard conservation strategy is to establish reserves, usually on publicly owned lands, with the goal of protecting nature by limiting human activities in the designated areas (Doremus 2010, 203-207).[36] The United States already has many protected areas where human activity is restricted to varying degrees, including wilderness areas, national monuments, wildlife refuges, national and state parks, and land subject to conservation easements. Biodiversity could be further protected by establishing additional reserves on existing publicly owned lands and in publicly managed ocean areas (see e.g., Karkkainen 1997, Wyman 2008, Johnson 2010). Designating new protected areas on publicly owned lands and in publicly managed oceans will not be politically easy given the human stakes in the federal public domain. Recall in this regard the battles over Forest Service roadless areas and wilderness area designations (Quarles 2010, 10817). But experience suggests that some progress might be made in designating new protected areas on federal lands if the designations are part of a compromise package that also allows or requires development in other areas of federal lands and waters (Johnson 2010, 10813).

Establishing additional reserves would reduce the momentousness of ESA listing decisions for unlisted and listed species. Creating new reserves would provide unlisted species in these areas with additional safeguards, and in so doing potentially avoid the need to list them and reduce the significance of not being listed under the ESA. The designation of new reserve areas inhabited by ESA listed species would make their listing under the ESA less vital for efforts to recover them. In an era of ongoing climate change, there is certainly an argument for establishing new protected habitats as the ecosystems in existing protected areas are altered.

To be sure, designating additional reserve areas on federal lands will not be a panacea for species. Many imperiled species also require habitat on privately owned lands (see Shaffer 2006, 291). Even species that may be helped through the designation of new federal reserves may require active assistance to survive, not just the designation of additional reserves (Scott et al. 2010; Doremus 2010, 206). The needs of species may evolve and the benefits they can gain from reserve lands may diminish, especially in an era of climate change. But these concerns are not insurmountable, although addressing them would require political will that often is hard to harness on behalf of conservation policy. They could be addressed to a degree by coupling the reserve strategy with more targeted incentives for private landowners to protect land for species (Scott et al. 2010, 95), establishing goals for managing reserves that include active management, and updating periodically the designating of reserved areas to reflect ongoing changes in climate and ecosystems. Overall, the reserve strategy likely is worth continuing to pursue notwithstanding the political obstacles. In addition to the benefits it offers for protecting biodiversity generally, it is a prophylactic measure that may reduce the need to list species under the ESA, and may make the ESA less significant for the recovery of currently listed species—and by extension indirectly reduce the significance of the pernicious effects of the interplay of science and policy in ESA listing decisions.

CONCLUSION

This chapter suggests setting to the side the efforts to separate science and policy in ESA listing decisions. The consequences of the existing science-policy interplay are problematic for science and policy. But given the obstacles to disentangling them it may be that the most we can do is to lessen the significance of these consequences as a by-product of pursuing other policy options that we should be undertaking in any event to improve protections for species and biodiversity generally. After almost 40 years, the ESA is showing its age and limits as a means of protecting species, not to mention biodiversity. Addressing these limits by amending the Act and/or supplementing it are two ideas worth pursuing not only because doing so may help reduce the importance of the politicization of science and the scientization of politics.

NOTES

1. The effort to delist the gray wolf has gone through several rounds of rulemaking and litigation and remains a work in progress. CRS 2011 provides a useful analysis with maps of the implications of the various rounds for wolf protection.

In 2003, FWS created three Distinct Population Segments (DPSs) for the wolf, which previously had been listed as endangered in the conterminous U.S., except in Minnesota, where it had been listed as threatened. It also downlisted two of these DPSs (the Eastern and Western DPSs) to threatened status, while leaving the third DPS (the Southwestern DPS) as endangered. The rule also delisted the gray wolf in fourteen southeastern states on the basis that the wolf should not have been listed in these states because they were not within the wolf's historical range (DOI FWS 4/2003a). Along with the rule, the FWS announced it intended to delist the Eastern and Western DPSs (DOI FWS 4/2003b; DOI FWS 4/2003c). Following successful judicial challenges to the rule by environmentalists, it was enjoined and vacated in *Defenders of Wildlife v. Secretary,* U.S. Dept., 354 F.Supp.2d 1156 (D. Or. 2005), and vacated and remanded for reconsideration in *National Wildlife Federation v. Norton,* 386 F.Supp.2d 553 (D.Vt. 2005).

In two separate rules in 2007 and 2008, the FWS designated and delisted a Great Lakes DPS and a Rocky Mountain DPS (DOI FWS 2/2007 [Great Lakes]; DOI FWS 2/2008 [Rocky Mountain]). The rule designating and delisting the Great Lakes DPS was vacated and remanded to the agency for reconsideration in *Humane Soc. of U.S. v. Kempthorne,* 579 F.Supp.2d 7 (D.D.C. 2008). The rule designating and delisting the Rocky Mountain DPS was preliminarily enjoined in *Defenders of Wildlife v. Hall,* 565 F.Supp. 2d 1160 (D. Mont. 2008). At the FWS's request, the court later vacated the rule and remanded it to the FWS for reconsideration (DOI FWS 12/2008).

In 2009, the FWS designated a Northern Rocky Mountain DPS and delisted it except in Wyoming (DOI FWS 4/2009a). In 2009, environmental groups sought, but did not win, a preliminary injunction to reinstate protections for the gray wolf after Idaho and Montana authorized wolf hunts following the delisting. However, the court ruling on the motion for a preliminary injunction indicated that the environmentalists likely would succeed on the merits in their challenge to the 2009 partial delisting. The partial delisting rule was vacated and set aside in *Defenders of Wildlife v. Salazar,* 729 F.Supp. 2d 1207 (D. Mont. 2010) and the FWS withdrew the rule (DOI FWS 10/2010). In 2011, in an indicative ruling, the federal district court that had vacated the rule refused to stay the vacatur after ten of the plaintiffs reached a settlement with the federal government that would have allowed the delisting to proceed for Idaho and Montana only, pending the resolution of an appeal. *Defenders of Wildlife v. Salazar,* 2011 WL1345670 (D. Mont. 2011).

Section 1713 of the Department of Defense and Full-Year Continuing Appropriations Act of 2011, P.L. 112-10 states "Before the end of the 60–day period beginning on the date of enactment of this Act, the Secretary of the Interior shall reissue the final rule published on April 2, 2009 et seq. [DOI FWS 4/2009a] without regard to any other provision of statute or regulation that applies to issuance of such rule. Such reissuance (including this section) shall not be subject to judicial review and shall not abrogate or otherwise have any effect on the order and judgment issued by the United States District Court for the District of Wyoming in Case Numbers 09–CV–118J and 09–CV–138J on November 18, 2010." The referenced decision is *Wyoming v. United States* DOI, 2010 U.S. Dist. LEXIS 122829 (D. Wyo., Nov. 18, 2010), in which the Wyoming Wolf Coalition challenged the continued listing of the wolf in Wyoming and the court held that the FWS's treatment of Wyoming was arbitrary and capricious and remanded to the FWS. The April 2, 2009 rule (DOI FWS 4/2009a) was republished on May 5, 2011 (DOI FWS 5/2011b). Congress's recent delisting is being challenged on constitutional grounds as a violation of the separation of powers. Complaint for Declaratory and Injunctive Relief, *Alliance for the Wild Rockies v. Salazar,* Case 9:11-cv-00070-DWM (D. Mont. May 5, 2011); Complaint for Declaratory and Injunctive Relief, *Center for Biological Diversity v. Ken Salazar,* Secretary of the Interior; Rowan Gould,

Acting U.S. Fish and Wildlife Service Director; and United States Fish and Wildlife Service, Case 9:11-cv-00071-DWM (D. Mont. May 5, 2011).

Also in 2009, the FWS designated and delisted a Great Lakes DPS (DOI FWS 4/2009b). The rule was challenged but the FWS reached a settlement with the plaintiffs and pursuant to the settlement the court remanded and vacated the 2009 rule and the FWS withdrew the rule (DOI FWS 9/2009). In 2011, on the same date it republished the rule designating and delisting the Northern Rocky Mountain DPS, the FWS proposed to designate and delist the Great Lakes DPS (DOI FWS 5/2011a).

2. I use the terms politics and policy interchangeably in the chapter, while recognizing they may have different connotations. For example, Pielke Jr. (2007, 31) defines policy as "a commitment to a particular course of action" and politics as "the process of bargaining, negotiation, and compromise that determines who gets what, when, and how."

3. Section 1533(b)(1)(A) of 16 U.S.C. states "The Secretary shall make determinations required by subsection (a)(1) solely on the basis of the best scientific and commercial data available to him. . . . " The referenced section 1533(a)(1) states "The Secretary shall by regulation promulgated in accordance with subsection (b) determine whether any species is an endangered or threatened species because of any of the following factors:

(A) the present or threatened destruction, modification, or curtailment of its habitat or range;
(B) overutilization for commercial, recreational, scientific, or educational purposes;
(C) disease or predation;
(D) the inadequacy of existing regulatory mechanisms; or
(E) other natural or manmade factors affecting its continued existence."

To quote Professor Doremus (1997, 1096), "[a]rguably the only 'determination'required by subsection (a)(1), and therefore the only determination subject to the strictly science mandate, is the viability status of groups already identified as species."

As a side note, it is doubtful that a listing agency could evaluate based solely on science "the inadequacy of existing regulatory mechanisms" as required by 1533(a)(1)(D). Rolhf 2004, 520–21.

4. Thank you to Professor Doremus for speaking to me about whether the interpretation she identified has been picked up in existing case law and for pointing me to the orca case.

5. However there are differences in the ways that the FWS and NOAA Fisheries deal with DPSs. See e.g., Doremus 2010, 198; Brosnan 2007, 995–96.

6. See Marris (2007, 252): "As Waples describes it, the distinct population segment combines a biological description of the relationships with a clear value judgement as to whether a population is important. 'You really need something besides science to decide where on that level you are going to focus,' he says."

7. The meaning of the phrase "in danger of extinction throughout all or a significant portion of its range" is currently under active consideration. After adverse judicial rulings, on May 4, 2011, the Solicitor of the Department of Interior withdrew a March 16, 2007 legal opinion defining the phrase. FWS and NOAA now intend to develop a joint policy on the meaning of the phrase. Federal Respondents' Motion For Voluntary Remand and Vacatur, *Center For Native Ecosystems v. Salazar*, Case 1:09-cv-01463-JLK, (D. Colo., May 13, 2011).

8. See note 1 for a chronology of the efforts to downlist and delist the gray wolf. The wolf delisting efforts have raised the question of whether the FWS can designate a DPS in order to delist it. See *Humane Soc. of U.S. v. Kempthorne*, 579 F.Supp.2d 7 (D.D.C. 2008); see also DOI FWS 4/2009a, 15129–15130.

9. The FWS's 1994 definition of recovery also has distributional and temporal components. "In defining a 'Recovered Wolf Population' we found 'in the northern Rockies a recovered wolf population is 10 breeding pairs of wolves in each of 3 areas for 3

successive years with some level of movement between areas'" [DOI FWS 4/2009a, 15131 (quoting Service 1994, p. 6-7)].

10. There were 1,651 Northern Rocky Mountain wolves in 2010. Rocky Mountain Wolf Recovery 2010 Interagency Annual Report.

11. For a technical definition of $N_{e,}$ see Traill et al. 2010, 3 (Box 1: Genetically effective population sizes).

12. For a more critical assessment of the FWS's surveys of biologists for the 1994 and 2001-2002 reviews, see NRDC 2008, 7–11.

13. Although the FWS does not mention public hostility to wolves in discussing the recovery goal, the rule discusses public hostility in analyzing the threats currently facing the wolf and likely to face it in the future, as required by 16 USC §1533(a)(1) (DOI FWS 4/2009a, 15175).

14. Pielke Jr. (2007, 14) describes "four idealized roles for scientists (and experts more generally) in decision-making": Pure Scientist, Issue Advocate, Science Arbiter and Honest Broker of Policy Alternatives.

15. See Crandall 2006, 250 ("scientific objectivity in conservation biology, in particular, is difficult (many of us become biologists *because of* our love of nature)"); Doremus 1997, 1149–1150; Biber 2011.

16. Pielke Jr. (2007, 94) indicates that "Stealth issue advocacy refers to situations in which one claims to be serving as a Pure Scientist or Science Arbiter, but instead is focused on reducing the scope of choice available to decision-makers, which is the defining characteristic of an Issue Advocate."

17. While the taxonomic status of the Preble's now is regarded as settled by the FWS in favor of it being a subspecies, the listing of the Preble's is still being debated.

In 2008, the FWS delisted the Preble's in Wyoming but left it listed as threatened in Colorado, after concluding that the Preble's is a taxonomically valid subspecies but that it is not threatened in Wyoming (DOI FWS 7/2008). The decision to delist the Preble's in Wyoming has been challenged on two fronts. Environmentalists have challenged the delisting of the Preble's in Wyoming and the listing of the Preble's in Colorado as threatened instead of endangered. See *Center for Native Ecosystems v. Salazar*, 711 F.Supp.2d 1267 (D. Colo. 2010). On May 13, 2011, in the suit brought by environmentalists, the federal government filed a motion for voluntary remand and vacatur of the 2008 partial delisting that would reinstate the 1998 listing decision while the federal government reconsiders the 2008 partial delisting. Federal Respondents' Motion For Voluntary Remand and Vacatur, *Center For Native Ecosystems v. Salazar*, Case 1:09-cv-01463-JLK, (D. Colo., May 13, 2011). The environmental groups have agreed to the remand and vacatur. Petitioners' Response to Federal Respondents' Motion For Voluntary Remand and Vacatur, *Center for Native Ecosystems v. Salazar*, Case 1:09-cv-01463-JLK (D. Colo., June 3, 2011). However there are three Wyoming intervenors (the State of Wyoming, the Wyoming Farm Bureau, and the Wyoming Stock Growers Association) that oppose the voluntary remand and vacatur. Respondent-Intervenor State of Wyoming's Response to Federal Respondents' Motion For Voluntary Remand and Vacatur, *Center for Native Ecosystems v. Salazar*, Case 1:09-cv-01463-JLK (D. Colo., June 3, 2011); Wyoming Farm Bureau Federation's and Wyoming Stock Growers Association's Response to Respondents' Motion For Voluntary Remand and Vacatur, *Center for Native Ecosystems v. Salazar*, Case 1:09-cv-01463-JLK (D. Colo., June 3, 2011). Separately, after the 2008 partial delisting, the Mountain States Legal Foundation, which represents the three intervenors in the suit brought by environmentalists, sued to oppose the continued listing of the Preble's in Colorado. First Amended Complaint/Petition for Review, *Mountain States Legal Foundation v. Dirk Kempthorne*, Case 03-cv-250-J (D. Wyo., October 17, 2008). This case apparently was dismissed.

18. See also Mannix 2009, 11 (similarly observing that the ESA makes relevant "scientific questions" that are "substantively unimportant"); Brosnan 2007, 993–98 (discussing the policy-driven science to which the ESA has given rise).

19. In this respect, the panel seemed to agree with REA, who argued in a response to KEA that "the differences between [their] conclusions is largely a function of basic conceptual and philosophical differences" (Ramey et al. 2007, 1). However, the panel did not attribute KEA and REA's different conclusions on whether the Preble's is a distinct subspecies entirely to philosophical differences. For an overview of the various reasons that the panel gives for the differences, see the executive summary of the panel's report. Arbogast et al. 2006, 3-4. For Ramey's concerns with the panel, see e.g., Congress 2007.

20. See Ramey et al. 2005, 332 ("greater molecular variance among than within subspecies . . . or . . . nearly complete reciprocal monophyly with respect to other subspecies"); King et al. 2006, 4333 ("significant phylogeographic separation of mtDNA alleles between subspecies (or populations), combined with congruent phylogeographic structure for nuclear loci").

21. Without endorsing either Ramey et al. (2005) or King et al. (2006), Brosi & Biber (2009) argue that Ramey et al. (2005) make a common inferential error, though one they later qualified, in indicating that the absence of evidence of differentiation proves the truth of their null hypothesis that the Preble's is not a distinct subspecies.

22. Carolan 2008a and Carolan 2008b, by a sociologist, highlight the differences in values underpinning the debate about whether the Preble's is a distinct subspecies, and I draw on Carolan's arguments in this section. For another interesting perspective on the debate in the 2000s about whether the Preble's is a distinct subspecies, see Crifasi 2007. As the Water Resources Administrator of the City of Boulder Open Space and Mountain Parks Department, Crifasi represented Boulder in developing a habitat conservation plan for the Preble's. Crifasi 2007, 529.

23. See Ramey et al. 2005, 329 ("If defensible data are lacking and a protected organism is not distinguishable with a high degree of certainty from neighbouring, non-threatened relatives, considerable financial and logistical conservation effort may be misallocated at the expense of other endangered organisms"); Ramey et al. 2007, 3 ("there are many endangered taxa and not enough resources to conserve them. If conservation effort is allocated to nondistinct or weakly differentiated populations, other more unique taxa (e.g. full species) will lose out."). See also Bottrill et al. (2008) (arguing for the application of triage principles to allocate limited conservation resources).

24. On the negative consequences of the scientization of politics and politicization of science for the listing process and science, see Wagner 1995, 1673–1688 (analyzing the negative consequences of the science charade in the regulation of toxic risks, but many of her arguments apply to the ESA); Doremus 1997, 1152–1153; Carolan 2008b, 462–63; Carolan 2008a, 949–50.

25. For other recommendations to facilitate more open, policy-oriented discussions of listing decisions see e.g., Rohlf (2004), Ruhl & Salzman (2006), Carolan 2008a, 2008b.

In 2004, a bill was introduced in Congress that overtly would have politicized the listing process with very different objectives than Professor Doremus's proposal. Among other changes, the Endangered Species Recovery Act of 2002, H.R. 4475, 108th Cong. (2004), would have dropped the requirement that listing decisions be based "solely" on the best available science, and required the approval of two of the three of the Secretary of the Interior, the Secretary of Transportation, and the Secretary of Defense to list a species.

26. Doremus 1997 (1132–1134) discusses other proposals to improve the scientific foundations of ESA listing decisions that presume listing decisions are scientific, including ideas of the National Research Council.

In the 2000s, many bills were introduced in Congress with the stated objective of improving the science used in ESA decision-making (including listing decisions). Many of these bills would have made it harder to list species under the Act and protect them, however. Many of these bills followed a standard pattern although they were not identical. Thus many of the bills required data collected in the field to support a listing decision; gave greater weight to empirical, field-tested, or peer-reviewed data in decision-making; and required peer review by an independent board of the science underlying listing and other decisions. See Sound Science for Endangered Species Act Planning Act of 2002, H.R. 4840, 107th Cong. (2002); Endangered Species Listing and Delisting Process Reform Act of 2003, S. 369, 108th Cong. (2003); Sound Science for Endangered Species Act Planning Act of 2003, H.R. 1662, 108th Cong. (2003); Scientifically Identifying the Need for Critical Habitat Act, H.R. 2602, 108th Cong. (2003); Sound Science for Endangered Species Act Planning Act of 2004, S. 2009, 10th Cong. (2004); Endangered Species Reform Act of 2007, 110th Cong. (2007). For a discussion of some of these bills, see Ruhl & Salzman 2006 (33–34). Bills intended to bring "sound science" to ESA decision-making date back to 1978 (Doremus 2004, 427–28).

One bill from the 2000s worth mentioning is the Threatened and Endangered Species Recovery Act of 2005, H.R. 3824, 109th Cong. (2005), which was passed by the House but not the Senate. This bill would have radically overhauled the ESA. Its listing-related changes included dropping the reference to commercial data, and requiring the issuance of regulations establishing criteria for determining which data is the best available science that would require the data to be empirical or peer reviewed and in compliance with guidance issued under the Data Quality Act. The bill also would have required that cost-benefit analyses be prepared concurrent with listing decisions. In addition, it would have directed that DPSs be listed sparingly. H.R. 3824 drew significant public attention. It was criticized in an editorial in *Science* which argued that it "would limit the use of well-tested population models for determining whether to add species to the endangered list" (Trombulak et al. 2006, 973).

The Information Quality Act (also called the Data Quality Act) was passed in 2000 in an appropriations bill. Its stated objective was to improve the information disseminated by agencies and to provide affected persons with administrative means to correct the information maintained and disseminated by agencies. Treasury and General Government Appropriations Act for Fiscal Year 2001, P.L. 106-554, §515. The Act seems to have had little impact on ESA listing decisions (Doremus 2005, 290). The FWS and the Department of Commerce maintain websites on their implementation of the Information Quality Act (DOI FWS, Office of External Affairs; DOC, Office of the Chief Information Officer).

27. Although not labeled this way, Ramey's other two main options for policy reform (phylogenetic ranking and conservation triage) are means of allocating limited conservation resources among listed groups, not means of deciding which groups to list. To use either approach to decide which groups to list, it would be necessary to define a threshold rank or score that species would need to be listed, a task that would resemble Ramey's proposal to define thresholds for listing.

28. Doremus (1997, 1139–1141) anticipates and responds to criticisms of her proposals for identifying the protected groups of populations.

29. These examples are drawn from the list of uncertainties relevant to the ESA in Doremus 2004, 438–39. Brosnan (2007, 999–1000) offers a more general list of sources of doubt in science.

30. See Doremus 2004, 419, arguing that "Taken as a whole, . . . the circumstances, language, and what legislative history there is suggest that the best available science mandate was generally intended to ensure objective, value-neutral decision making by specially trained experts." Biber 2011, 50–55 offers another interpretation of the best available science mandate as a precommitment device to enlist a discipline sympathetic to the ESA's conservation goals.

31. My discussion of the ESA's track record in helping species draws on Wyman 2008, 494–99 and the sources cited therein.

32. By regulation, the FWS (but not NOAA Fisheries) has extended the prohibition on takings to threatened species under its jurisdiction except when a species-specific rule is passed that varies the prohibition. 50 C.F.R. §17.31 & 17.71. NOAA Fisheries does not presumptively apply the takings prohibition to threatened species. Rohlf 2004, 530–31 note 183 and accompanying text; Quarles 2010, 10815, note 3.

33. For a historical perspective on FWS efforts to manage pressures to list species and designate critical habitat, see Rohlf 2004, 495–96, 526–28.

34. In Wyman (2008, 518) I suggest that the goal for species could be set at something more realistic. But here I consider that possibility in discussing Quarles's reform proposal.

35. See also Rohlf 2004 (508–509) (proposing that federal agencies be legally required to implement recovery plans).

36. The ESA itself makes use of the preserve strategy, for example through the designation of critical habitat, which can regarded as "(sharply) limited preserves" (Doremus 2010, 208).

REFERENCES

Agapow, Paul-Michael et al. (2004). "The Impact of Species Concept on Biodiversity Studies." *Quarterly Review of Biology* 79(2): 161–79.

Arbogast, B. S., J. P. Dumbacher, and S. J. Steppan (2006). *Evaluation of Scientific Information Regarding Preble's Meadow Jumping Mouse.* Portland, OR: Sustainable Ecosystems Institute.

Bergstrom, Bradley J., et al. (2009). "The Northern Rocky Mountain Gray Wolf is Not Yet Recovered." *BioScience* 59(11): 991–99.

Biber, Eric (2011). "Which Science? Whose Science? How Scientific Disciplines Can Shape Environmental Law." Unpublished manuscript.

Biber, Eric, and Berry Brosi (2010). "Officious Intermeddlers or Citizen Experts? Petitions and Public Production of Information in Environmental Law." *UCLA Law Review* 58(2): 321–400.

Bottrill, Madeleine, et al. (2008). "Is Conservation Triage Just Smart Decision Making?" *Trends in Ecology and Evolution* 23(12): 649–54.

Brosi, Berry J., and Eric G. Biber (2009). "Statistical Inference, Type II Error and Decision Making Under the US Endangered Species Act." *Frontiers in Ecology and the Environment* 7(9): 487–94.

Brosnan, Deborah M. (2007). "Science, Law, and the Environment: The Making of a Modern Discipline." *Environmental Law* 37: 987–1006.

Carolan, Michael S. (2008a). "Is It a Distinct Subspecies? Preble's Mouse and the 'Best Available Science' Mandate of the Endangered Species Act." *Society & Natural Resources* 21(10): 944-51.

Carolan, Michael S. (2008b). "The Politics in Environmental Science: The Endangered Species Act and the Preble's Mouse Controversy." *Environmental Politics* 17(3): 449–65.

Congressional Research Service (2011, April 21). "The Gray Wolf and the Endangered Species Act: a Brief Legal History." CRS Report R41730. [CRS 2011].

Crandall, K. A. (2006). "Letter from the Editor: Advocacy Dressed Up as Scientific Critique." *Animal Conservation* 9: 250–51.

Crifasi, Robert R. (2007). "A Subspecies No More? A Mouse, its Unstable Taxonomy, and Western Riparian Resource Conflict." *Cultural Geographies* 14: 511–35.

Cronin, Matthew A. (2007). "The Preble's Meadow Jumping Mouse: Subjective Subspecies, Advocacy and Management." *Animal Conservation* 10: 159–61.

Cronin, Matthew A. (2006). "A Proposal to Eliminate Redundant Terminology for Intra-Species Groups." *Wildlife Society Bulletin* 34(1): 237–41.

Cronin, Matthew A. (1997). "Systematics, Taxonomy, and the Endangered Species Act: The Example of the California Gnatcatcher." Wildlife Society Bulletin, 25 (3) 661–66.

Davis, Frank W. (2006). "Renewing the Conservation Commitment." In The Endangered Species Act at Thirty: Renewing the Conservation Promise, vol. 1, ed. Dale D. Goble, et al., 296–306. Washington, Covelo & London: Island Press.

Department of Commerce. Office of the Chief Information Officer (2010). "Good Guidance—Information Quality." ocio.os.doc.gov/ITPolicyandPrograms/Information_Quality/index.htm [DOC, Office of the Chief Information Officer]

Department of Commerce. National Oceanic and Atmospheric Administration (1991). "Policy on Applying the Definition of Species Under the Endangered Species Act to Pacific Salmon," 56 Fed. Reg. 58612 (Nov. 20, 1991) [DOC NOAA 1991].

Department of the Interior. Fish and Wildlife Service. Department of Commerce. National Oceanic and Atmospheric Administration (1996). "Policy Regarding the Recognition of Distinct Vertebrate Population Segments Under the Endangered Species Act," 61 Fed.Reg. 4722 (Feb. 7, 1996) [DOI FWS & DOC NOAA 1996].

Department of the Interior. Fish and Wildlife Service (2005). Endangered and Threatened Wildlife and Plants. "12-Month Finding on a Petition to Delist the Preble's Meadow Jumping Mouse and Proposed Delisting of the Preble's Meadow Jumping Mouse," 70 Fed. Reg. 5405 (Feb. 2, 2005) (to be codified at 50 C.F.R. pt. 17)[DOI FWS 2/2005].

Department of the Interior, Fish and Wildlife Service (2008). Endangered and Threatened Wildlife and Plants: "Final Rule Designating the Northern Rocky Mountain Population of Gray Wolves as a Distinct Population Segment and Removing This Distinct Population Segment From the Federal List of Endangered and Threatened Wildlife." 73 Fed. Reg. 10514 (February 27, 2008) (to be codified at 50 C.F.R. pt. 17) [DOI FWS 2/2008].

Department of the Interior, Fish and Wildlife Service (2007). Endangered and Threatened Wildlife and Plants: "Final Rule Designating the Western Great Lakes Populations of Gray Wolves as a Distinct Population Segment; Removing the Western Great Lakes Distinct Population Segment of the Gray Wolf From the List of Endangered and Threatened Wildlife." 72 Fed. Reg. 6052 (February 8, 2007) (to be codified at 50 C.F.R. pt. 17) [DOI FWS 2/2007].

Department of the Interior, Fish and Wildlife Service (2008). Endangered and Threatened Wildlife and Plants: "Final Rule to Amend the Listing for the Preble's Meadow Jumping Mouse to Specify Over What Portion of its Range the Subspecies is Threatened," 73 Fed. Reg. 39790 (July 10, 2008) (to be codified at 50 C.F.R. pt. 17) [DOI FWS 7/2008].

Department of the Interior, Fish and Wildlife Service (2009). Endangered and Threatened Wildlife and Plants: "Final Rule to Identify the Northern Rocky Mountain Population of Gray Wolf as a Distinct Population Segment and to Revise the List of Endangered and Threatened Wildlife," 74 Fed. Reg. 15123 (April 2, 2009) (to be codified at 50 C.F.R. pt. 17) [DOI FWS 4/2009a].

Department of the Interior, Fish and Wildlife Service (2009). Endangered and Threatened Wildlife and Plants: "Final Rule to Identify the Western Great Lakes Populations of Gray Wolf as a Distinct Population Segment and to Revise the List of Endangered and Threatened Wildlife," 74 Fed. Reg. 15070 (April 2, 2009) (to be codified at 50 C.F.R. pt. 17) [DOI FWS 4/2009b].

Department of the Interior, Fish and Wildlife Service (2003). Endangered and Threatened Wildlife and Plants: "Final Rule to Reclassify and Remove the Gray Wolf From the List of Endangered and Threatened Wildlife in Portions of the Conterminous United States; Establishment of Two Special Regulations for Threatened Gray Wolves," 68 Fed. Reg. 15804 (April 1, 2003) (to be codified at 50 C.F.R. pt. 17) [DOI FWS 4/2003a].

Department of the Interior, Fish and Wildlife Service (2011). Endangered and Threatened Wildlife and Plants: "Proposed Rule to Revise the List of Endangered and Threatened Wildlife for the Gray Wolf (Canis lupus) in the Eastern United States, Initiation of Status Reviews for the Gray Wolf and for the Eastern Wolf (Canis Lycaon)," 76 Fed. Reg. 26086 (May 5, 2011)(to be codified at 50 C.F.R. pt. 17) [DOI FWS 5/2011a].

Department of the Interior, Fish and Wildlife Service (1994). "The Reintroduction of Gray Wolves to Yellowstone National Park and Central Idaho. Final Environmental

Impact Statement." www.fws.gov/mountain-prairie/species/mammals/wolf/EIS_1994.pdf.

Department of the Interior, Fish and Wildlife Service (2010). Endangered and Threatened Wildlife and Plants: "Reinstatement of Protections for the Gray Wolf in the Northern Rocky Mountains in Compliance With a Court Order," 75 Fed. Reg. 65574 (October 26, 2010) (to be codified at 50 C.F.R. pt. 17) [DOI FWS 10/2010].

Department of the Interior, Fish and Wildlife Service (2008). Endangered and Threatened Wildlife and Plants: "Reinstatement of Protections for the Gray Wolf in the Western Great Lakes and Northern Rocky Mountains in Compliance With Court Orders," 73 Fed. Reg. 75356 (December 11, 2008)(to be codified at 50 C.F.R. pt. 17) [DOI FWS 12/2008].

Department of the Interior, Fish and Wildlife Service (2009). Endangered and Threatened Wildlife and Plants: "Reinstatement of Protections for the Gray Wolf in the Western Great Lakes in Compliance With Settlement Agreement and Court Order," 74 Fed. Reg. 47483 (September 16, 2009)(to be codified at 50 C.F.R. pt. 17) [DOI FWS 9/2009].

Department of the Interior, Fish and Wildlife Service (2011). Endangered and Threatened Wildlife and Plants: "Reissuance of Final Rule to Identify the Northern Rocky Mountain Population of Gray Wolf as a Distinct Population Segment and to Revise the List of Endangered and Threatened Wildlife," 76 Fed. Reg. 25590 (May 5, 2011) (to be codified at 50 C.F.R. pt. 17) [DOI FWS 5/2011b].

Department of the Interior, Fish and Wildlife Service (2003). Endangered and Threatened Wildlife and Plants: "Removing the Eastern Distinct Population Segment of Gray Wolf From the List of Endangered and Threatened Wildlife," 68 Fed. Reg. 15876 (April 1, 2003) (to be codified at 50 C.F.R. pt. 17) [DOI FWS 4/2003b].

Department of the Interior, Fish and Wildlife Service (2003). Endangered and Threatened Wildlife and Plants: "Removing the Western Distinct Population Segment of Gray Wolf From the List of Endangered and Threatened Wildlife," 68 Fed. Reg. 15879 (April 1, 2003) (to be codified at 50 C.F.R. pt. 17) [DOI FWS 4/2003c].

Department of the Interior, Fish and Wildlife Service (2009). Endangered and Threatened Wildlife and Plants: "Revised Critical Habitat for the Preble's Meadow Jumping Mouse in Colorado," 74 Fed. Reg. 52006 (Oct. 8, 2009) (to be codified at 50 C.F.R. pt. 17).[DOI FWS 9/2009b].

Department of the Interior, Fish and Wildlife Service (2012). Office of External Affairs. "Information Quality." www.fws.gov/informationquality/ [DOI FWS, Office of External Affairs]

Department of the Interior, Fish and Wildlife Service et al. (2010). "Rocky Mountain Wolf Recovery 2010 Interagency Annual Report." www.fws.gov/mountain-prairie/species/mammals/wolf/annualrpt10/FINAL_2010_Northern_Rockies_Summary_and_Background_3_9_11.pdf

Department of the Interior, Fish and Wildlife Service (2011a). "Species Reports, Environmental Conservation Online System, Summary of Listed Species Listed Populations1 and Recovery Plans" as of Wed, 08 Jun 2011 20:26:42 GMT. ecos.fws.gov/tess_public/pub/Boxscore.do. [DOI FWS 2011a]

Department of the Interior, Fish and Wildlife Service (2011b). "Species Reports, Environmental Conservation Online System, Delisting Report." www.fws.gov/ecos/ajax/tess_public/DelistingReport.do. [DOI FWS 2011b]

Doremus, Holly (2010). "The Endangered Species Act: Static Law Meets Dynamic World." *Washington University Journal of Law & Policy* 32: 175–235.

Doremus, Holly (1997). "Listing Decisions Under the Endangered Species Act: Why Better Science Isn't Always Better Policy." *Washington University Law Quarterly* 75 (3): 1029–1153.

Doremus, Holly (2004). "The Purposes, Effects, and Future of the Endangered Species Act's Best Available Science Mandate." *Environmental Law* 34: 397–450.

Doremus, Holly (2005). "Science Plays Defense: Natural Resource Management in the Bush Administration." *Ecology Law Quarterly* 32: 249–306.

General Accounting Office (2003). Endangered Species: Fish and Wildlife Service Uses Best Available Science to Make Listing Decisions, but Additional Guidance Needed for Critical Habitat Designations. GAO-03-803. [GAO2003].

Gifford, Krishna (2007). "Measuring Recovery Success." Endangered Species Bulletin. 4: 4.

Goble, Dale D. (2009). "The Endangered Species Act: What We Talk About When We Talk About Recovery." *Natural Resources Journal* 49: 1-44.

Goble, Dale D. (2010). "Recovery," in *Endangered Species Act: Law, Policy and Perspectives*, edited by Donald C. Baur and Wm. Robert Irvin, 70-103. Chicago: American Bar Association.

Harbine, Jenny K. (2009). "Gray Wolves in the Northern Rockies Again Staring Down the Barrel at Hostile State Management." *Ecology Law Currents* 36: 195–204.

Holthouse, David (2005). "Building a Better Mousetrap: Biologist Rob Ramey Isn't Afraid of Taking Risks but Taking on Environmentalists may be his Riskiest Move." *Denver West WorldNews*, January 20.

Houck, Oliver (1993). "The Endangered Species Act and its Implementation by the U.S. Departments of Interior and Commerce." *University of Colorado Law Review* 64(2): 277-370.

Irvin, Wm. Robert (2010). "Comment on Rethinking the ESA to Reflect Human Dominion Over Nature." *Environmental Law and Policy Annual Review* 40: 10809-10811.

Issac, Nick J. B. (2004). "Taxonomic Inflation: Its Influence on Macroecology and Conservation." *Trends in Ecology and Evolution*, 19(9): 464–69.

Johnson, Richard P. (2010). "Above All, Try Something: Two Small Steps Forward for Endangered Species." *Environmental Law and Policy Annual Review* 40: 10812–10814.

Karkkainen, Bradley C. (1997). "Biodiversity and Land." *Cornell Law Review* 83: 1–104.

King, Tim et al. (2006). "Comprehensive Genetic Analyses reveal Evolutionary Distinction of a Mouse (Zapus hudsonious preblei) Proposed for Delisting from the US Endangered Species Act." *Molecular Ecology* 15: 4331–4359.

Male, Timothy D. & Michael J. Bates (2005). "Measuring Progress in US Endangered Species Conservation." *Ecology Letters* 8: 986.

Mannix, Brian F. (2009). "Dumb Queues and Not-So-Bright Lines: the Use and Abuse of Science in the Endangered Species Act." American Enterprise Institute for Public Policy Research. Paper presented at the "Rebuilding the Ark: New Perspectives on ESA Reform" conference.

Maron, Dina Fine (2011). "Endangered Species: Judge Halts Historic Agreement Designed to Reduce Litigation Backlog." *E&E News*, May 5. www.eenews.net/climatewire/2011/05/18/archive.

Marris, Emma (2007). "The Species and the Specious." *Nature* 446: 250–53.

Martin, A. (2006). Letter to the Editor: "Advocacy Dressed Up As Science: Response to Ramey et al." *Animal Conservation* 9: 248–49.

Mayden, R. L. (1997). "A Hierarchy of Species Concepts: the Denouement in the Sage of the Species Problem," in *Species*, edited by M. F. Claridge, H. A. Dawah, and M. R. Wilson, 381–424. London: Chapman & Hall.

Meazell, Emily Hammond (2010–2011). "Super Deference, the Science Obsession, and Judicial Review as Translation of Agency Science." *Michigan Law Review* 109: 733–84.

Natural Resource Defense Council (2008). "Public Comment on Designation of the Northern Rocky Mountain Population of Gray Wolf as a Distinct Population Segment and Removing This Distinct Population Segment From the Federal List of Endangered and Threatened Wildlife." November 28.

New York Times (2011). Editorial: "A Hole in the Endangered Species Act." *New York Times*, April 21.

NOAA Fisheries. Office of Protected Resources. Endangered Species Act (ESA) (2012). www.nmfs.noaa.gov/pr/laws/esa/.

Pielke, Jr., Roger A (2007). *The Honest Broker: Making Sense of Science in Policy and Politics*. Cambridge UK: Cambridge University Press.

Quarles, Steven P. (2010). "Wyman's Rethinking the ESA: Right Diagnosis, Wrong Remedies." *Environmental Law and Policy Annual Review* 40: 10815–10818.

Ramey, Rob R. II et al. (2005). "Genetic Relatedness of the Preble's Meadow Jumping Mouse (Zapus hudsonius preblei) to Nearby Subspecies of Z. Hudsonius as Inferred from Variation in Cranial Morphology, Mitochondrial DNA and Microsatellite DNA: Implication for Taxonomy and Conservation." *Animal Conservation* 8: 329–46.

Ramey, Rob R. II et al. (2007). "How King et al. (2006) Define an Evolutionary Distinction of a Mouse Subspecies: a Response." *Molecular Ecology* 16: 3518–3521.

Ramey, Rob R. II et al. (2006). Response to Vignieri et al (2006): "Should Hypothesis Testing or Selective Post Hoc Interpretation of Results Guide the Allocation of Conservation Effort?" *Animal Conservation* 9: 244–47.

Ramey, Rob R. II (2010). "On the Origin of Specious Species." Paper presented at Institutions and Incentive in Regulatory Science conference, June 7.

Rohlf, Daniel (2004). Section 4 of the Endangered Species Act: "Top Ten Issues for the Next Thirty Years." Environmental Law 34: 483–553.

Ruhl, J. B. and James Salzman (2006). "In Defense of Regulatory Peer Review." *Washington University Law Review* 84(1): 1–61.

Scott, J. Michael, et al. (2006). "By the Numbers," in *The Endangered Species Act at Thirty: Renewing the Conservation Promise* vol. 1, edited by Dale D. Goble et al., 16–35. Washington, Covelo & London: Island Press.

Scott, J. Michael, et al. (2010). "Conservation-Reliant Species and the Future of Conservation," *Conservation Letters* 3: 91–97.

Shaffer, Mark L. (2006). "Proactive Habitat Conservation," in *The Endangered Species Act at Thirty: Renewing the Conservation Promise*, vol. 1, edited by Dale D. Goble et al., 286–95. Washington, Covelo & London: Island Press.

Taylor, Phil. 2011. "Wolves: Groups Sue Over Delisting Rider, say Congress Violated Constitution." *E&E News*, May 5. www.eenews.net/eenewspm/print/2011/05/05/3.

Traill, Lochran W., et al. (2010). "Pragmatic Population Viability Targets in a Rapidly Changing World." *Biological Conservation* 143: 28–34.

Trombulak, Stephen C. et al. (2006). "Science as Smoke Screen." *Science* 312: 973.

U.S. Congress, House Committee on Resources (2004). Bill to Amend the Endangered Species Act of 1973 to Reform the Process for Designating Critical Habitat Under That Act: Hearing on H.R. 2933 Before the House Committee on Resources, 108th Cong., 2nd Sess.

U.S. Congress, House Committee on Natural Resources (2007). Crisis of Confidence: the Political Influence of the Bush Administration on Agency Science and Decision-making: Oversight Hearing Before the House Committee on Natural Resources (statement of Rob Ramey II), 110th Cong. 1st Sess.

Wagner, Wendy E. (1995). "The Science Charade in Toxic Risk Regulation." *Columbia Law Review* 95(7): 1613–1723.

Woody, Todd (2011). "Wildlife at Risk Face Long Line at U.S. Agency." *New York Times*, April 21.

Wyman, Katrina M. (2008). "Rethinking the ESA to Reflect Human Dominion Over Nature." *NYU Environmental Law Journal* 17(1): 490–528.

Zuckerman, L. (2010). "Protections for Wolf, Grizzly, Might be Lifted." *Washington Post*, December 5.

III

Reforming the Role of Science in Environmental, Health, and Safety Regulation

SIX

Reconciling the Scientific and Regulatory Timetables

James W. Conrad Jr.[1]

One of the fundamental and chronic challenges afflicting governmental use of science for policy purposes is the conflict between the essentially continuous evolution of scientific knowledge and the discrete nature of government decision making. Two related features of science help produce this conflict: (i) the truism that the scientific process never reaches an end; and (ii) the more specific fact that the validity and importance of an individual study are generally not ascertainable immediately, but depend on the extent to which it can be replicated and extended by further work (and thus gain adherents). Government, on the other hand, inherently must make decisions at given points in time, and at those times "freezes" the state of the relevant science in a legislative or administrative act. And due to the inertia and delays inherent in bureaucratic processes, that frozen state may persist for years or decades.[2]

This conflict can emerge in either of two situations. Sometimes government needs or wants to act, but the science isn't "ready." In these cases, government must either (i) delay acting, or else (ii) lock in a scientific conclusion that is not adequately supported or at risk of becoming quickly obsolete. Other times, the science is ready, but government is unwilling or unable to act. In these cases, government is by default perpetuating the status quo, which could be a previously "frozen" and now outdated scientific position. In any of these cases, depending on the facts, the result can potentially be harm to protected or regulated interests.

In some cases, the temporal conflict between science and government is irreducibly real. More often, however, the asserted conflict is a cover for self-interested behavior by players in the regulatory process—above all, by government itself. This chapter will discuss several examples of these situations in the context of chemical regulation. The chapter will then discuss ways that claims of conflict between scientific and regulatory timetables can be better assessed and true conflicts identified. Where those conflicts exist, the chapter will discuss how they can be ameliorated by scientific and administrative mechanisms.

The examples below illustrate the dispiriting extent to which regulatory agencies are lagging indicators of scientific progress and require continual oversight by others to spur them in the direction of new thinking. In the area of chemical risk assessment, this fact raises significant concerns about how objectively or effectively agencies will make use of new "omics" approaches that have the potential to revolutionize how we assess chemical risks (see Conrad 2011, §§ 4:4.1, 4:16 [explaining toxicogenomics and related fields and the toxicity pathway approach to assessing chemical hazards]). Adoption by agencies of the ameliorative mechanisms discussed below would go a long way toward alleviating those concerns.

I. BACKGROUND—THE TEMPORAL CONFLICT AND THE RESPONSIBILITY OF PLAYERS IN THE REGULATORY PROCESS

"[T]he scientific process of seeking the truth, by design and to its credit, has no natural endpoint" (NRC 2009, 47). New discoveries regularly upend settled theories, and even where an experiment or discovery answers a question, it inevitably spawns new questions to be explored. Another essential feature of the scientific enterprise is that it is not refereed externally. Instead, the validity of a given work is ultimately determined by consensus within the relevant scientific community, depending principally on whether studies can be replicated or extended. But this process takes time, and disputes about the correctness of theories and findings can rage for decades. It has been argued that most published results will in time turn out not to be correct, but rather the product of selection bias among experimenters, publication bias by journals, reanalysis of data in search of a significant result, and the inherently provisional assurance offered by statistical significance tests. Thus, at any point in time, the "state of the science" on a given issue will have some degree of uncertainty, and much of what is regarded plausible, if not likely, will eventually turn out to be wrong.[3] These traits are particularly true of "regulatory science," the interdisciplinary and policy-generated scientific fields that underlie much of health and environmental risk assessment (Jasanoff 1990, 76–80).

Regulatory agencies define legal obligations through adjudication or rulemaking. Also, and especially relevant in science-dependent areas, they regularly make determinations that do not have independent legal force, but are later incorporated in legally binding actions. In any of these situations, however, decisions are made and become effective on specific dates. Typically, they remain unchanged until they are amended or superseded at some later date. However incomplete or evolving it is, the state of the science supporting a regulatory decision will effectively be cast in concrete at the time the decision is finalized. Thus, while science may move along continuous trajectories of various shapes, regulatory decisions based upon it will at best approximate it with discrete steps, like an Etch-a-Sketch drawing.[4]

In any given case, the science supporting a regulatory decision will be more or less "ready," which I define as either (i) generally agreed upon, among disinterested (or uninvolved) members of the relevant scientific community, as sufficient for the purposes of the government action, or at least (ii) not likely to change dramatically in the near term. Whether *interested* persons might agree will depend on things like how central the scientific question is to the decision, how wrong the science might be, and, above all, the stakes at issue: how costly compliance will be, the asserted benefits of the action, and how precedent-setting it will be. The greater the magnitude of any of these factors, the more agreement among scientists or stability will be required before stakeholders will concede that the science is ready. Indeed, for issues of huge cost and controversy, like the existence and causes of climate change, the science will have been ready a long time ago for some people, but will never be ready for others.

Players in the regulatory process will inevitably make whatever ready/not ready arguments suit their interest. Sometimes these claims will be legitimate—it is certainly not the case that self-interested claims are always wrong. But frequently such claims will be bogus, and an effective regulatory system will continuously need to be alert to the potential for such claims and try to reject them.

To some extent, industry and nongovernmental organizations (NGOs) can be forgiven for advocating for their interests—interest groups will inherently err on the side of advocacy, each expecting that the other will push back. Indeed, that is the principal purpose of many of the groups most active in the regulatory process. The federal government has a unique responsibility, though, as the most legitimate representative of the public interest, and as beneficiary of all stakeholders' tax payments. We can expect it *not* to make bogus claims—exactly as we expect the Justice Department to have as its goal producing justice, not winning in every case.[5] As shown below, however, regulatory agencies are among the biggest offenders.

II. BOGUS INVOCATIONS OF THE CONFLICT

All players in the regulatory process will, from time to time, strategically seek to accelerate or delay action by spurious reference to the state of the science. The invocations basically take two forms: stalling for time and rushing to judgment. Below I describe multiple examples of federal agencies engaging in each of these practices in the area of chemical regulation. I then discuss why the government seems so attracted to doing what it ought not do.

At the outset, though, a quick primer on the fundamentals of chemical regulation. In 1983, the National Academy of Sciences (NAS) published its landmark book *Risk Assessment in the Federal Government: Managing the Process*, which codified the now universally accepted paradigm that the risk from a substance is a function of two things: its inherent hazard (how "toxic" it is) and the extent to which a person is exposed to it (the "dose") (NRC 1983, 20). The hazard of a substance is typically expressed in terms of a dose that, over a lifetime of exposure, is without appreciable risk (in the case of non-carcinogens) or would not increase one's background risk of cancer by 1 in a 100,000 or more. Such hazards are typically estimated using toxicological studies in which lab animals are fed high doses of the substance, although where available assessors will also use epidemiological data (e.g., studies of exposed workers) and mechanistic data (i.e., evidence of how the substance acts on the body in the process of being absorbed and metabolized). Knowledge regarding such toxicological "modes of action" has particularly increased in recent years, supporting the development of complex models.

Several government initiatives generate chemical hazard assessments, the most influential of which is clearly EPA's "Integrated Risk Information System" or "IRIS" program (see EPA, IRIS Basic Information). This centralized database of chemical toxicity assessments is relied upon by all EPA offices except the pesticide program, and also by states and foreign countries. Another prominent assessment process is the periodically updated Report on Carcinogens (RoC), a congressionally mandated document issued by the National Toxicology Program (NTP), an interagency effort overseen by the National Institute of Environmental Health Sciences (NIEHS) within the Department of Health and Human Services (HHS). The RoC identifies chemicals that NTP regards as "known . . . or . . . reasonably anticipated to be carcinogens" (see 42 U.S.C. § 241[b][4][A]).

A. Stalling for Time

Sometimes the science on a given chemical hazard really is ready, and additional study is unlikely to materially shift the weight of evidence. Frequently, however, interested groups will (i) call for or commission studies that will not really produce decision-relevant science (for example, because they involve tangential or inconsequential issues); (ii) convene expert panels or hire individual experts to criticize government views or scientific consensus or to create validation for minority views; or (iii) simply criticize studies supporting the other side's position as "junk science."

Industry groups are most typically charged, in David Michaels' memorable phrase, with "manufacturing uncertainty" in this way (Michaels 2008, x). This is not surprising, really; where enormous sums of money are at stake, it is cost-justifiable to spend some fraction of it on science advocacy. How far down this road a company or industry may be willing to go will be a function of factors like (i) how profitable a product or activity is; (ii) how much capital has been sunk into it; (iii) how difficult it is for the company to diversify or switch products; and (iv) how little the company cares about public opinion. For example, compare tobacco (highly profitable and widely reviled) or petroleum refining (huge sunk costs; limited ability to diversify; only slightly more popular) with a large, diversified consumer product manufacturer (may be easily able to redeploy assets, and its reputation likely its greatest asset).

What others perceive as stalling for time is not always or even mostly dishonest—in most cases, the scientific staff of entities accused of stalling are sincerely convinced of their correctness. All scientific and technical fields have a range of respectable views (OMB 2005), and people on one end of the range tend to be more inclined to work for industry. And as these deeply engaged scientists look more closely at the relevant issues, the more likely they are to perceive, and view as important, questions that others would overlook or regard as unimportant.[6]

NGOs will also stall for time when it suits them—most typically, in permitting or deregulatory contexts. Again, this usually reflects sincere activity—NGO scientists tend to come from the other end of the range just mentioned.

But federal agencies also stall for time, routinely. In part this may be because their scientific staff tend to come from the NGO end of the range. Also, senior career staff can have great influence, and can often outflank or outlast political appointees with whom they disagree. But the issue is much more complex than that. Below I present a series of prominent case studies involving EPA primarily, but also NIEHS. As they show, federal agencies regularly delay assessments for two apparent purposes: (i) to give them time to generate data, arguments, or expert reviews that will rebut industry-supported data or methodological work, and (ii) to allow them to continue to regulate the relevant chemical using the agency's existing, long-ago "frozen" health assessments. The result, in the words of the National Academy of Sciences' "Silver Book," is that "[t]he regulatory risk assessment process is bogged down" [NRC 2009, 3 (citing the examples of formaldehyde and dioxin discussed infra and of trichloroethylene)].

1. Formaldehyde

As I write this, in July 2011, it seems likely that this year's blockbuster NAS report will be the one it issued in April on the draft health assessment of formaldehyde prepared by EPA's IRIS program (NRC 2011). The NAS report details a host of "recurring methodologic problems" afflicting the draft (3), but the most remarkable fact it notes early on is the length of time the process has taken EPA (14):

1990/1991—EPA publishes initial noncancer and cancer assessments.

January 1998—EPA initiates reassessment.

October 2004—EPA begins developing draft reassessment document.

June 2010—Draft released for NAS and public review.

NAS does not essay to explain the reasons for these extensive delays, but the shortcomings of the report that they identify gives a clue. Bear in mind that the formaldehyde industry has been devoting substantial sums to toxicological and modeling work over this entire timeframe, in hopes of influencing both EPA's IRIS assessment and a similar cancer assessment being conducted by NTP, and has been providing its work product to both agencies. And yet, the NAS found that:

EPA discussed but did not use a computational fluid dynamics model developed with industry support, even though it would be "useful" (5);

EPA did not use biologically based dose response (BBDR) models developed with industry support, even though it said one them was among "the best-developed BBDR models to date"; instead, it rejected the models on the basis of manipulations that the NAS found "extreme" and concluded "should not have been used" to reject them (5). (The committee also expressed "concern that EPA directed substantial effort toward refuting many of the assumptions and conclusions of the . . . models rather than trying to fill the data gaps that were clearly articulated by the models."[7]);

In several cases, EPA relied on very old studies (1987 and 1990) to the exclusion of newer, equally or more reliable, industry supported studies (6).

EPA rejected animal studies showing adverse effects on nasal tissues even though "[t]he animal studies offer one of the most extensive datasets on an inhaled chemical, and EPA should use the[m]" (6–7).

Regarding the key cancer assessment, NAS stated:

> [T]he draft IRIS assessment does not provide a clear framework for causal determinations. As a result, the conclusions appear to be based on a subjective view of the overall data, and the absence of a causal framework for these cancers is particularly problematic given the inconsistencies in the epidemiologic data, the weak animal data, and the

> lack of mechanistic data. Although EPA provided an exhaustive description of the studies and speculated extensively on possible modes of action, the causal determinations are not supported by the narrative provided in the draft IRIS assessment (8).

NAS summarized:

> In general, the committee found that the draft was not prepared in a consistent fashion; it lacks clear links to an underlying conceptual framework; and it does not contain sufficient documentation on methods and criteria for identifying evidence from epidemiologic and experimental studies, for critically evaluating individual studies, for assessing the weight of evidence, and for selecting studies for derivation of the RfCs [i.e., concentrations in air that can be inhaled without appreciable lifetime risk] and unit risk estimates. (3–4)

The most logical interpretation of the foregoing findings is that EPA has devoted much of the past thirteen years not to determining and articulating what constitutes the best available science on formaldehyde, but rather to developing arguments about why it should rely on the studies that show the highest hazard, many of which are now decades old, and why it should reject studies, datasets, and models that tend to show much less hazard. Additional time seems to have been consumed in drafting a "roughly 1,000 page draft" containing "long descriptions of the studies" but "little beyond a brief introductory chapter . . . on the methods for conducting the assessment" (3).

2. Dioxin

The NAS formaldehyde report begins by remarking that the problems with the formaldehyde assessment "are similar to those which have been reported over the last decade by other [NAS] committees tasked with reviewing EPA's IRIS assessments for other chemicals" (3), and it devotes most of the last chapter to "suggestions for improvements in the IRIS development process. . . . " (112). The formaldehyde report specifically highlighted EPA's dioxin reassessment (18), and with good reason.

The dioxin reassessment has actually been going on even longer than the formaldehyde reassessment, since 1991. After two earlier drafts were reviewed by EPA's Science Advisory Board (SAB; a standing body of external scientists), EPA produced a third draft in 2003, which was referred to the NAS in 2004. NAS issued a report in 2006 making two fundamental criticisms of that draft.

The first involved EPA's "no-threshold" or "linear" assumption for the carcinogenicity of dioxin. Animal cancer studies use the highest doses that will not kill the animals outright because studies using realistic environmental exposures would require prohibitively costly and inhumane numbers of animals before statistically significant effects could be detected. This creates the question of how to extrapolate from test doses to environmental doses. Risk assessors in the 1970s adopted the presumption that there was no threshold to cancer's mode of action and that any dose, down to a single molecule, could potentially cause cancer. A formidable body of evidence has since emerged that many cancer-causing substances actually act through a threshold mechanism, and hence a graph of dose against effect would not be a straight line starting at the intersection of the two axes. As illustrated repeatedly below, however, agencies have been extremely reluctant to accept these mechanisms.

Concluding that "EPA's decision to rely solely on a default linear model lacked adequate scientific support," NAS urged EPA to provide risk estimates based on a threshold or nonlinear model as well (NRC 2006, 5). NAS also found that EPA "did not adequately quantify the uncertainty associated with" its cancer estimate, and encouraged EPA to include a quantitative analysis of uncertainty in its next draft (6).

EPA took four years to respond to NAS, and yet its 2010 response (not a revised draft but a supplement to the 2003 draft) argued that only the linear cancer model could be justified. It also did not present a quantitative uncertainty analysis. Not surprisingly, EPA referred this draft to the SAB, not NAS. In a draft report issued earlier this year, the SAB allowed as how the report was "clear" and "logical," but "identified [the same two] deficiencies" that NAS had complained of (EPA SAB 5/2011, i). The SAB found that "the Report did not respond adequately to the NAS recommendation to adopt both linear and nonlinear methods of extrapolation" (ii), explaining that "[i]t is, in fact, the fundamentally nonlinear nature of the dose-response for receptor mediated processes that underlies the conviction of a large segment of the scientific community that a nonlinear approach should be preferred for the risk assessment for dioxin" (35). It also did "not agree with the position taken in the Report that [a quantitative uncertainty] analysis is unfeasible" (iii). Finally, echoing NAS's general observations, the SAB also noted that "[t]he Report is long and dense with a considerable amount of jargon and in some places it is quite repetitive. These features, while a necessity of this type of document, at times detract from clarity or make the EPA's logic difficult to discern" (12).

As with formaldehyde, industries that generate dioxin as an unintended by-product of manufacturing or combustion have been actively supporting work to better understand the mechanisms of dioxin toxicity and have been advocating favorable work—some conducted by NTP—for use by EPA (see, e.g., AHR Case Study, 2010). It is difficult to avoid the conclusion that much of what EPA has been doing over the past two decades, and particularly from 2003–2010, has been to rebut this work, buttress its own longstanding conclusions, and generally obscure the logic behind its actions. In a telling comment, an SAB member was paraphrased as saying at a public meeting that EPA "seems to focus on presenting evidence that supports use of the linear model rather than a balanced evaluation of the alternatives" (Hogue 2010 [paraphrasing Harvey Clewell]).

EPA has indicated that it now will respond to NAS and SAB's recommendations (Hogue 2010). Meanwhile, EPA has been relying on its original dioxin IRIS file, which dates to 1984, to propose "Preliminary Remediation Goals" for soil cleanup (in the process, rejecting a massive University of Michigan biomonitoring project showing that soils are not significant sources of dioxin exposure [ACC Comments, 2009]). John Doull, coauthor of the leading toxicology treatise, has opined that:

> EPA is premature in setting soil standards when it has not even set standards for human exposure to dioxin. As a result of the process EPA has set into motion based on narrowly-selected science, we are being led to believe, contrary to a large body of studies, that any exposure to dioxin is harmful. [EPA's standards] call into question the body of scientific studies we know about, change the course of how we will answer questions of potential harm in ways that are not validated by science and do so without any explanation of what is to be gained for public health. (Doull & Waddell 2010)

EPA has also this year released final dioxin "toxicity equivalency factors" for the various dioxin congeners, in the face of criticism from industry for acting before it completes its reassessment (Hogue 2010, 32).

EPA rules and voluntary action have reduced total U.S. anthropogenic dioxin emissions 92 percent since 1987 (ACC 2010), and dioxin levels in human blood dropped seven-fold from 1972–1995 (Hites 2011, 20). EPA's dioxin reassessment, stalled throughout most of this time, has had essentially nothing to do with this "almost incidental reduction" (20).

3. Methanol and Related Chemicals

EPA spent most of a decade evaluating the carcinogenicity of methanol, based on work supplied to it in 2002 by the European Ramazzini Foundation, located in Italy. The Methanol Institute, which has sponsored extensive toxicological work on methanol, contended throughout that time that EPA should not rely on the Ramazzini work because the lab does not follow study guidelines observed by EPA, NTP, or the Organisation for Economic Cooperation and Development (OECD)—for example, it allows animals to die of old age rather than euthanizing them at a predetermined age. Ramazzini's interpretation of slides has long been viewed skeptically by many because of its insistence that aspartame causes cancer in rats, contrary to multiple other major studies, and because of its refusal to cooperate with the FDA on the issue. (Ramazzini would only share partial data with the FDA, but based on it, the FDA "identified significant shortcomings in the design, conduct, reporting, and interpretation of th[e] study" [FDA 2007]).

In October 2009, EPA announced that it had asked the SAB to review a forthcoming draft assessment of methanol (EPA 10/2009), ultimately released in January 2010, that was based on Ramazzini data. NTP had already decided to send a team of pathologists to Ramazzini to review a sample of their slides, and EPA also worked with NTP so that the review would include the methanol slides. To EPA's surprise, however, in June 2010 the NTP delegation reported substantial discrepancies between Ramazzini's reports and the underlying slides,[8] and so EPA was forced to cancel the SAB review and place a hold on work to update the IRIS files for methanol and three related compounds (EPA 6/2010).

After ten months—in April 2011—EPA reinitiated work on the non-cancer assessment for methanol and all assessments of the other three chemicals because, as EPA explained, those could be performed without reliance on Ramazzini data. EPA also announced that it had formed a Pathology Working Group to be jointly administered by EPA and NIEHS to fully address questions about Ramazzini's work—something that industry groups had long been calling for. Finally, EPA noted that two already-completed assessments (for vinyl chloride and 1,1 dichloroethylene) had also relied on Ramazzini work; it stated that it will reassess these assessments after gets the results of the Pathology Working Group (EPA 4/2011). It will be a significant development if, as a result of that effort, EPA finally stops trying to rely on Ramazzini's work and thereafter subjects all data to a consistent, rigorous process of scientific quality review and strength of evidence evaluation, as recommended by the NAS Formaldehyde Report (NRC 2011, 114–15).

4. Inorganic Arsenic

EPA issued a pair of documents in 2005 addressing the carcinogenicity of inorganic arsenic (iAs) and asked its SAB to review them. The EPA documents concluded that iAs causes bladder and lung cancer, based on Taiwanese drinking water studies, with a linear mode of action. The SAB's report, issued in 2007, said that reliance on these studies and the linear assumption was reasonable given the state of knowledge "at this time" (EPA SAB 2007, 7), adding:

> However, the Panel recognized limitations to the[Taiwan]ese data, and that there is some evidence on iAs from animal toxicology, pharmacokinetics, and pharmacodynamics research, that suggests other than a linear bladder cancer dose-response. The Panel urged the Agency to consider other epidemiologic studies from the U.S. and other countries, utilizing a uniform set of evaluative criteria. The Panel also recommended sensitivity analyses be conducted to account for human variability. (2)

Almost three years later (February 2010), EPA issued a 575-page draft reassessment that was notable primarily for its continued reliance on the

Taiwanese studies and its avoidance of the SAB's recommendations (EPA, External Review Draft, 2010). Five members of the 2007 SAB panel wrote EPA to complain that:

- "In the five years that have ensued since our review of the 2005 draft document," better epidemiology studies have emerged, "but the draft did not present a review of each study conducted by systematic consistent application of the uniform performance criteria" called for in 2007. "For example, smoking status documentation is still inconsistent . . . even though a large body of epidemiological literature has documented smoking behavior as a co-carcinogenic or synergistic factor in bladder and lung cancer causation. Smoking status information is also lacking in the Taiwan based studies utilized to derive the cancer slope factor" (Comments of Aaron Barchowsky et al. 2010, 1–2).
- "Similarly, an extensive compilation of *in vitro* study references with brief descriptions is presented . . . [but] no coherent critical integration and evaluation of these data is presented. . . . Meanwhile, other scientists have successfully evaluated, integrated and published critical reviews of this data base" (2).

Originally, EPA did not plan further SAB review or public comment on the 2010 draft, but acceded under pressure to an "expedited and focused" peer review using a new SAB panel containing none of the 2007 reviewers. It announced a sixty-day public comment period, but then said that that only comments received within thirty-five days would be provided to the reviewers—and then failed to provide any of them to the reviewers.

The new SAB panel noted that it "was not asked to conduct a full peer review of the assessment, including EPA's calculation of the cancer risk estimate" (EPA SAB 3/2011, 1), and also expressed "concern" that EPA's revised draft did not consider any studies published after 2007 (thus excluding more than 300 peer-reviewed publications) (8). While the SAB endorsed EPA's use of the Taiwanese studies, it cautioned that "[EPA] needs to state more clearly the set of criteria that EPA used in evaluating and presenting the studies" (2). It also approved use of a linear assumption, although it noted that the resulting "upper-bound cancer risk estimate for arsenic . . . is of significant public health concern. The SAB suggests that EPA discuss, possibly in other EPA risk assessment documents, how the estimated risks for arsenic should be interpreted in light of current estimated bladder and lung cancer incidence for the U.S. population." As clarified by an SAB member in a public teleconference,[9] the SAB panel was tactfully saying that if arsenic is as carcinogenic as EPA is hypothesizing, EPA needs to perform a "reality check" to explain whether it really accounts for all or most of current levels of these cancers, as the assessment would lead one to conclude.[10] Finally, SAB members complained about "the elephant in the room here"—the narrowness of the charge questions.[11]

Admittedly, the foregoing assessments are some of the most politically controversial that EPA is conducting and involve large volumes of scientific information. Nonetheless, the clear pattern that emerges is EPA stalling for years, not because the best available science isn't "ready," but because EPA does not want to use it. Instead, EPA appears primarily to be spending its time developing lengthy documents explaining why the new methodological work isn't ready or why EPA's preferred studies are still the best. EPA also appears to be using bodies like the NAS and SAB strategically to obtain expert validation of its position, sometimes successfully and sometimes not.

5. NIEHS

As noted earlier, EPA is not alone in the practice. NIEHS has been the chief proponent within the federal government of the notion that certain substances can have serious adverse effects on developing embryos through endocrine-modulated modes of action, even (or *only*) at very low doses. This theory is certainly worth exploration, although it has been extensively challenged in its most controversial example, bisphenol A (BPA) (Sharpe 2010, 1–4; Goodman et al. 2006, 387–457). However, several of NIEHS's approaches suggest that NIEHS may be stalling for time, attempting to buttress the credibility or numerosity of studies finding low-dose effects against contrary literature. The clearest example is the fact that NIEHS funded a three-day workshop in November 2006 at which thirty-eight invited experts discussed the state of the science on BPA's endocrine effects, resulting in the "Chapel Hill Consensus Statement" (vom Saal et al. 2007, 131–38). The only problem is that all of these scientists already endorsed the concept—NIEHS made no attempt to include scientists who are skeptical that BPA causes adverse effects at low doses. NIEHS's bias in this regard is also supported by its support for many years of studies that involved administration of BPA by injection, a practice that has been roundly criticized by the FDA and others as involving a biologically irrelevant exposure route when one's concern is assessing the safety of oral ingestion (FDA 2008, 9). After enough criticism, NIEHS announced that it would no longer fund such work without special justification (Butterworth 2009).

B. Rushing to Judgment

The obverse of stalling for time is rushing to judgment: that is, where the science is not ready to be used for regulatory purposes, but a player in the process insists that the scientific questions have already been more than adequately answered and action is warranted post-haste. Again, any player in the process, including industry, can make this argument where it suits the player's agenda, but the practice is most typically engaged in by NGOs who are convinced that more than enough evidence already exists to ban or restrict particular chemicals or classes of chemicals. For example:

> Scientific evidence increasingly shows that BPA poses serious health risks. . . . Our children should not be used as guinea pigs by chemical companies while we wait, potentially for many years, for additional scientific evidence to show this chemical is harmful to our health. (Office of U.S. Senator Dianne Feinstein, 2011)

Sometimes the advocacy is pretty obviously just that. In September 2009, a long list of researchers wrote the FDA Commissioner (copying NIEHS) complaining about FDA's plan to spend $10 million over five years studying BPA. They found it "troubling that the FDA is proposing to spend such a large amount of money on such a well-researched chemical." Citing the Chapel Hill Consensus Statement referenced above, they said "there is sufficient research and independent review available for the agency to make a decision as to whether, as the law dictates, there is 'reasonable certainty' that this chemical is 'not harmful'" (Letter from Thomas Zoeller et al. to Margaret Hamburg, 2009).

But only five weeks later, NIEHS announced that it had awarded seven of the letter's signatories (along with three others) $30 million—three times as much money—to study the very same chemical. Far from concluding that "there is sufficient research" to decide if BPA should be restricted, NIEHS declared that, "[w]hile recent assessments by authorities in the United States, Canada, Europe, and Japan agree that current food contact uses of BPA are safe, these assessments have identified the need to address data gaps" (NIEHS 2011). The grantees' applications clearly were pending even as they were denouncing FDA. NIEHS's press release had its Director Linda Birnbaum saying the grants were "to provide the best science we can."[12] Given the public statements of those seven PIs (principal investigators), however, one is tempted to conclude that the "best science" was that which will cast further suspicion on BPA—as opposed to what FDA was funding, which evidently had the potential to show that BPA did not warrant being banned.

As with stalling for time, one would expect or at least hope that the federal government would refrain from rushing to judgment. As the following cases studies illustrate, while it engages in this practice less often, it is a repeat offender here, too.[13]

1. Hexavalent Chromium

EPA compiled a literature search regarding hexavalent chromium (Cr_6) in 2008, and in that year indicated that it would not issue a revised risk assessment until 2012.[14] In November 2009, EPA published a notice asking interested persons to submit data on Cr_6 by January 2010 (EPA 11/2009). But only five months later, EPA had already completed a draft assessment and was seeking interagency comment on it (EPA, Hexavalent Chromium External Review Draft, 2010). The draft relied on a 2008 NTP rat study that involved doses 4–5 orders of magnitude greater than could possibly occur in the environment, and took no account of the recognized fact that Cr_6 is reduced to trivalent chromium (Cr_3) in the digestive tract. Interested industry groups—none of which actually manufacture chromium—initiated two projects: (i) a study of the mode of action by which Cr_6 might cause cancer (which includes investigating both the threshold and nonthreshold hypotheses), and (ii) a physiologically-based pharmacokinetic (PBPK) model that estimates the concentrations of Cr_6 and Cr_3 throughout the metabolic process. Both of these are slotted to be done in 2011 and will result in multiple articles to be submitted to peer-reviewed journals. Most members of a peer review panel organized by EPA to review its draft assessment "expressed serious concerns" with the draft's assumption of a nonthreshold cancer mode of action (Association of Metropolitan Water Agencies 2011, 2–3). Noting that the industry projects directly address the data gaps that EPA identified in its draft assessment, the reviewers asked EPA to take these new studies into account because they "may have a tremendous impact on [EPA's] final conclusions" (Rizzuto 2011, A-9). Notwithstanding these facts, EPA has retained a fourth quarter 2011 completion date for its assessment.[15] Thus, it appears that EPA will not consider the new work in developing its final assessment.

2. Atrazine

EPA reregistered this pesticide in 2006. Ordinarily, it would not have reviewed it again until 2013. In response to sustained adverse publicity by an NGO and the news media, however, EPA launched a reassessment in 2009. In late 2010, EPA's "FIFRA Scientific Advisory Panel" (a standing body of external scientists addressed solely to pesticide issues) met to comment on the studies on which EPA proposes to rely to assess risk of atrazine in drinking water. EPA's attempt to show increased risk from the chemical led to criticism from the panel, which according to press reports (Kovski 2010, A-12):

- "[E]xpressed strong misgivings about the statistical strength of epidemiological studies that have attempted to detect links between atrazine exposure and various non-cancer health effects, notably reproductive problems." (EPA "had itself conceded that the statistical power of the studies was not adequate to stand alone, but the agency still wanted the advisers' opinions on the merits of those studies."); and

- "[S]howed discomfort over the extent to which laboratory studies with rats could be translated into possible effects in humans. . . . [It's j]ust another area of uncertainty, unfortunately," remarked one panelist.

3. Styrene

As with stalling for time, EPA is not alone in rushing to judgment; the NTP has also practiced the phenomenon as part of its periodically updated Report on Carcinogens (RoC), a congressionally mandated document identifying chemicals that NTP regards as known or reasonably anticipated to be carcinogens (see 42 U.S.C. § 241[b][4][A]). The RoC evaluation process for a given chemical begins with a "Background Document" containing an exhaustive description of published toxicological and epidemiological studies on the chemical, as well as mechanistic and other relevant data. Until the most recent RoC, the Background Document remained unchanged as it went through a series of reviews by staff of NTP and other agencies and peer review by an expert panel. Comments, no matter how critical, were simply appended to the "frozen" background document.

After years of criticism, NTP piloted a new process for the twelfth edition of the RoC under which the Background Document was revised as it moved through the staff and expert review process. However, that process still resists input by external parties that might change the fundamental findings that undergird the conclusions that NTP ultimately reaches. A prime example is the case of styrene, one of the chemicals in the twelfth RoC. Under the new pilot process, an Expert Panel is tasked with (i) peer reviewing the draft Background Document, (ii) recommending whether to list the substance, and (iii) providing the scientific justification for its listing recommendation (NTP 2011). NTP seeks public comments on the work of the Expert Panel, but only as to the second and third tasks. The Panel's peer review comments on the draft Background Document, and any related changes to the document that the Panel recommends, are, therefore, not subject to public comment. The consequence of this omission is that NTP can and does finalize the Background Document *before* it receives comments on the Expert Panel's work.

The fundamental flaw of this approach was evident in the styrene review. The Expert Panel recommended numerous changes to the draft Background Document so that it would support the Panel's recommended listing and justification. A variety of industry groups filed extensive comments criticizing the Expert Panel's recommended changes to the Background Report. Unfortunately, NTP finalized the document before the due date for the comments, adopting all of the Expert Panel's recommendations, and refused to consider revising it in light of those comments. NTP's rush to judgment resulted in a Background Document that was "frozen" before NTP could even learn of deficiencies in it.

C. Why is Government So Prone to Stalling for Time and Rushing to Judgment?

Several features intrinsic to government lead it to regularly engage in the sorts of tactics just discussed. Most obvious are bureaucratic factors like the difficulties and delays inherent in government contracting, and budgetary uncertainties, which would tend to make agencies want to rely on work they have already done or possess, rather than to take on new work or run the risk of not having an adequate budget later. These issues are worth addressing, but they do not appear to be the principal problems. Rather, the real drivers seem to be:

- *The desire to claim success in protecting the public.* This is another inherent feature of regulatory agencies, which after all were created for that purpose. The flip side of this phenomenon is the desire to avoid criticism for caving in or becoming captive to regulated entities. Both of these biases are exacerbated by, and anticipate, pressure from Congress and the news media to be more protective and avoid being compromised. Agencies are typically praised for acting protectively and attacked for not doing so, and so tend to act accordingly.
- *The biases and agendas of agency staff.* On balance, people who work for regulated agencies tend to believe that the activities or substances they regulate do in fact threaten the public interest and need to be restricted. And working in such a culture tends to reinforce such beliefs.
- *The fact that it can.* Agency chemical risk assessment processes tend to have arisen as closed mechanisms where the agencies had complete discretion. For example, EPA's IRIS database began as an internal mechanism to promote consensus among the (non-pesticide) program offices at EPA on health risk values for particular chemicals (so that, for example, the waste and water offices did not set different cancer values for the same substance) (EPA, IRIS Basic Information). As the database became more and more influential, stakeholders succeeded in getting EPA to open the process up by giving notice of impending reviews and soliciting information for them. But IRIS values are not subject to notice and comment and are not "appealable" by anyone. The NTP's Report on Carcinogens has been deemed judicially reviewable, but NTP has always been upheld (Tozzi v. DHHS, 271 F.3d 30, 31 (D.C. Cir. 2001); Synthetic Organic Chemical Manufacturers Ass'n v. DHHS, 720 F. Supp. 1244 (W.D. La. 1989)). Thus, the agencies administering these and simi-

lar programs know that, in the end, they are not accountable to anyone for the way they run them.

Supporters of greater regulation argue that the biases described in the first two bullets above are appropriate, since the statutes that regulatory agencies implement typically have protective or precautionary biases. But such biases infect the entire risk assessment process, skewing it at every step, making it difficult for risk managers to obtain an objective perspective of risk or to separate risk assessment from risk management, both of which are important in a legal system that requires agencies to act rationally (see 5 U.S.C. § 706[2][A] [authorizing agency action to be struck down when arbitrary and capricious]; Pub. L. No. 106–554, § 515; OMB 2002 [requiring information upon which agencies rely to have "objectivity"]), and to rely on objective scientific information. These biases also pose a chronic hazard of their own, which is controlling the risks that the agency is charged with worrying about at expense of not controlling (or worsening) others. And the lack of accountability described in the third bullet only exacerbates these problems.

Proponents of more protective agency action also complain of a contrary bias, in that health regulatory agencies like FDA and EPA typically only base regulatory actions on results from tests conducted using internationally established test guidelines. (For example, FDA recently described the status of the science regarding BPA by comparing "[s]tudies employing standardized toxicity tests used globally for regulatory decision making" with "recent studies using novel approaches and different endpoints [that] evaluated developmental or behavioral effects that are not typically assessed in standardized tests" (FDA 2010)).

In most cases, however, there are multiple guideline-compliant studies on which an agency can rely, and by latching onto the one that indicates adverse health effects in some sex of some species of lab animal at a low dose, agencies like EPA and NTP have had little theoretical difficulty justifying conclusions about cancer causation or other chronic health effects at doses that frequently are equivalent to background levels.

These critics have a point, however, which is that regulatory agencies typically are lagging indicators of scientific progress. Moreover, because staff at "program" (i.e., standard-setting) offices like EPA's Office of Pesticide Programs are legally obligated to engage with outside stakeholders, they are forced to confront new theoretical developments. Staff at the EPA IRIS and NTP Report on Carcinogens programs, however, have no regulatory obligations or other requirements to consider commenters' views, and still operate in a highly isolated fashion with a high degree of discretion. As a result, despite the dramatic growth in knowledge and theory regarding toxicology, these programs function with a worldview not fundamentally changed from the 1970s, when federal agencies first formulated policies on how to assess chemical risks.

In 2007, the National Academy of Sciences issued *Toxicity Testing in the 21st Century: A Vision and a Strategy* (that year's blockbuster), which described how:

> Advances in toxicogenomics, bioinformatics, systems biology, epigenetics, and computational toxicology could transform toxicity testing from a system based on whole-animal testing [and] generate more robust data on the potential risks to humans posed by exposure to environmental agents[, thus] offer[ing] the prospect of improved risk-based regulatory decisions and possibly greater public confidence in and acceptance of the decisions. (NRC 2007)

To be useful for regulatory purposes, as the NAS explains, screens and tests that detect "perturbations to toxicity pathways" using these methods will have to be validated as reliably corresponding to disease in humans (NRC 2007, 171–77). The conservative and unilateral tendencies of the government agencies discussed above, however, raise significant concerns about how objectively or effectively they will employ these new approaches. Given the repeated willingness of these agencies to devote more time to fighting methodological advances than to making them work, regulated entities can be forgiven for wondering whether these agencies will concentrate on interpreting and applying these new approaches in similarly self-serving ways. The next part of this chapter describes partial solutions to the conflict between the regulatory and scientific timetables, some of which may ultimately be superseded by these new approaches, but which if used in the meantime would promote confidence in government and its ability to accept scientific progress evenhandedly.

III. NEEDED SOLUTIONS

At EPA's request, the National Academy of Sciences in January 2011 initiated a study that may consider, among other things, "[a]pproaches to more effectively deal with the inherent tensions among research, risk assessment, and regulatory timeframes" (NRC 2011). The issue discussed in this chapter is thus one of recognized concern. This final part of the chapter offers three solutions, again in the context of chemical assessment: (i) use of consensus criteria and frameworks methods for judging whether science is ready; (ii) where it isn't, collaborative processes for generating additional information; and (iii) administrative mechanisms to allow agencies to act when science is ready and yet retain flexibility to change.

A. Science Bounds the Usefulness of Any Solutions

It is important to reiterate that the tensions between scientific and regulatory timetables are partly intractable, and thus not all conflict claims are bogus. While Congress and EPA appear to have been largely successful in "forcing technology" in the area of pollution control, "forcing science" appears to be a more chancy proposition. Forty years in, the War on Cancer shows no sign of ending soon, as cancer has turned out to be multiple diseases without any single solution (see generally Muckherjee 2010). In 1996, Congress gave EPA three years to establish a program of validated screens and tests for assessing the potential of various chemicals to be endocrine disruptors. Environmental groups sued three years later when the program had not been established. Over a decade later, EPA has yet to complete the evaluation of a single chemical. While a recent EPA Inspector General blamed the delays in part on management shortcomings, it also noted "the many complexities in methods development and validation of Tier 1 assays"; that is, the difficulties in understanding endocrine-modulating mechanisms of action and in efficiently measuring a chemical's tendency to trigger them (EPA 5/2011). A consulting toxicologist deeply involved in the process since its inception was more blunt:

> It must be appreciated that neither EPA nor any scientist or scientific body is able to dictate the results of scientific research and the timeframe on which it will yield useable results. To my knowledge and from my perspective, EPA worked as rapidly as it could, taking advantage of cooperative efforts by the OECD and other international organizations, to conduct validation experiments and to adjust the experimental plan as necessary based on results of the studies as they were obtained. Some results enabled rapid progress; other results dictated abandoning initial approaches and evaluating alternatives instead. Scientific results simply cannot be forcefit to meet a predetermined schedule. (Testimony of Christopher J. Borgert 2010, 3)

But the problem is not completely, or even mostly intractable: for most chemicals and endpoints of concern, the types of data needed are understood, and it is far from clear that public health threats are so potentially serious as to require action before the data can be generated. The keys are to make optimal use of existing data, to generate missing data efficiently and timely, and to act promptly and flexibly throughout the process.

B. Agreed-Upon Methods for Evaluating Science

Recall the conclusion of the NAS's Formaldehyde report:

> In general, the committee found that the draft was not prepared in a consistent fashion; it lacks clear links to an underlying conceptual framework; and it does not contain sufficient documentation on meth-

> ods and criteria for identifying evidence from epidemiologic and experimental studies, for critically evaluating individual studies, for assessing the weight of evidence, and for selecting studies for derivation of the RfCs and unit risk estimates. (NRC 2011)

These are "recurring . . . problems," moreover, as the NAS noted and other case studies above have illustrated. In a great many cases, particularly those of well-studied chemicals like formaldehyde, dioxin, and arsenic, an enormous body of toxicological, epidemiological, and mechanistic data exists. The challenge facing agencies is to evaluate that data transparently, rigorously, and even-handedly according to frameworks that the scientific community have developed and concur in. These frameworks exist both for evaluating the reliability of a single study and for assessing weight of evidence of multiple studies. EPA's own policies in some cases call for their use, but as shown above those policies are often honored in the breach. Use of such frameworks is central to assuring that stakeholders have confidence in each other's analyses and to minimizing disputes over interpretation.

1. Reliability Criteria

The most widely agreed framework for evaluating the reliability of a single study is the Klimisch criteria, developed to assist in evaluating animal test data for the European Union's REACH program. These criteria give the highest grade of reliability ("reliable without restrictions") to studies that were conducted following internationally accepted test guidelines and Good Laboratory Practice (GLP) standards that mandate creation and government access to complete experimental documentation (Klimisch et al. 1996, 1–5). Comparable standards exist for Good Epidemiological Practice (GEP) (International Epidemiological Association 2007).

2. Mode of Action Frameworks

As the field of toxicology has evolved more sophisticated understandings of how exactly particular agents cause disease, scientists from EPA, NTP, and similar agencies around the world have collaborated with academic and industry scientists through organizations like the International Programme on Chemical Safety (IPCS) and the International Life Sciences Institute to develop structured frameworks or decision trees to harmonize and promote rigor and transparency of analysis of mode of action, the best-known of which is the IPCS Human Relevance Framework (Boobis 2006). Given the centrality of this issue to the assessment of virtually every suspected carcinogen, probably no single reform would reduce the controversy associated with those assessments than consistent and clear application of this framework.

3. Causation Frameworks

The most momentous leap in chemical safety assessment is concluding that an association shows causation and not just correlation. Some of the leading references on this topic are not frameworks but lists of criteria. The most famous of these is Sir Austin Bradford Hill's considerations for inferring causality from epidemiological data (Hill 1965); a more recent set oriented toward toxicological data was issued under the auspices of the Annapolis Institute in 2001 (Gray et al. 2001). More recently, toxicologists collaborating internationally have developed explicit frameworks for conducting "evidence-based toxicology" (Hoffmann & Hartung 2006; Guzelian et al. 2005). Again, careful and transparent use of these systems would minimize and focus what are now routine interpretive disputes. It should help prevent health values being based on "narrowly selected science . . . contrary to a large body of studies" (see Doull & Waddell 2010).

C. Filling Data Gaps, Credibly and in Time

In some cases, agencies will conclude that dispositive science does not yet exist. The key change that is needed in such cases is for agencies and stakeholders to seek to agree upon methods for generating the additional information. The essential elements of this change are ways to improve the likelihood that the new information (i) will be sufficient (and be seen as such by interested stakeholders); and (ii) will be generated in time to be used.

1. Collaboration on Substance

A fundamental shortcoming of EPA's IRIS program and similar agency systems is inadequate engagement of interested outside stakeholders at the outset of the work. As noted earlier, EPA publishes annual notices soliciting studies regarding chemicals to be evaluated that year, but EPA does not sit down with them to discuss which studies are the most important and why, what modes of action are proposed to be operative, what the most significant data gaps are, and what sorts of studies and experimental outcomes would be sufficient to fill those gaps. For most chemicals, such a "state of the science" symposium could probably be concluded in a day. While well-studied but controversial chemicals might require a multi-day session, it seems indisputable that bringing proponents of different views in a room to question and challenge each other would inform agency participants far better and more quickly than the current process, and thus should expedite and simplify their subsequent work process. Ideally, agency staff would engage in such symposia as questioners and moderators to ensure that issues are fully ventilated.

To the extent existing data arguably answer the most important questions, such symposia could seek to generate consensus, or at least narrow the range of dispute, regarding how frameworks for assessing mode of action or causation would be applied to those data. To the extent gaps arguably exist, the same dialogue could seek to make the same progress regarding exactly what new studies ought to be done. This dialogue could conceivably involve agreement on new, non-guideline methods that could be collaboratively developed.

It is likely that such symposia would be subject to the Federal Advisory Committee Act (FACA), which would somewhat heighten their associated administrative burdens. This could be handled by having existing FACA bodies (e.g., EPA's SAB) run them, or by outsourcing them to contractors (provided that EPA staff continued to actively participate). In any event, it still seems likely that the delays associated with state-of-the-science symposia at the outset of an assessment would be far less than the years of delay that we have seen result when other FACA bodies (like the NAS) review a completed (but flawed) draft assessment and send EPA back to do the work over. The cost savings would almost certainly be similarly proportional.

EPA and HHS currently allow industrial and other stakeholders to have input into peer reviews of draft IRIS files and RoC Background Documents and Expert Panel Reports, and NTP even responds to comments in final RoCs. But these opportunities seem to have been created more to allow these agencies to say they have provided due process than as mechanisms by which the agencies expect to be truly educated. In a setting where there is no administrative or judicial review, a "response to comments" document really serves no useful purpose beyond transparency, and many of these are so conclusory and opaque that they do not even serve that purpose. What *would* be useful is a more participatory, symposium-like meeting of the peer reviewers, interested stakeholders and agency staff. This process would include:

- Adequate time for stakeholders to review the draft document and prepare useful comments;
- Adequate time for the reviewers to review those comments before the meeting; and, most important;
- Opportunity for stakeholders to engage in dialogue with the reviewers, at least. If agency staff were also required to be present, that would enable panel members to ask them about questions raised by commenters.

These end-of-the-process reforms would probably cause less dramatic improvement in chemical assessment than "state of the science" symposia at the beginning of the process, but they would also be easier to integrate into existing peer review mechanisms.

2. Coordination of Schedules

Where a "state of the science" symposium produced agreement (or least a compelling contention) that a data gap existed and a particular study would fill it, the obvious next step in an improved chemical assessment process would be for the agency to determine whether manufacturers of that substance (or others) would be willing to undertake that study within an adequate but aggressive timetable. Where such willingness exists, the agency ought to schedule the assessment to be able to take that work into account (again, barring some public health reason for acting more precipitously). This would avoid the circumstance that so often happens now, where an agency launches ahead and completes an assessment even as important relevant research is ongoing, research which when ultimately completed often results in criticism of the now-outdated assessment.

NTP ostensibly includes a chemical in an upcoming RoC only when its staff conclude that existing data are sufficient for applying the listing criteria; if not, the nomination is returned with a recommendation that new data be generated (see NTP 2005, 2). That decision-making process is not currently public and dialogic, however. An annual "state of the science" event could enable NTP to begin scheduling chemicals for future reports based upon commitments to conduct research—although it should also include a willingness to defer them from current reports based on a similar commitment, where stakeholders could show that existing data actually were not sufficient to apply the listing criteria. EPA's process for determining what chemicals it will update in the IRIS database involves only internal considerations that do not include whether an interested stakeholder has initiated (or offered to initiate) a relevant study (see EPA, IRIS Frequent Questions). There is no reason why EPA could not do this, however.

EPA staff do communicate informally with sponsors of chemical testing regarding ongoing work, and in practice have occasionally timed assessments to make use of such work. There is no formal process for this coordination, however, and EPA never commits to use the work if completed according to plan. Such agreement would be crucial to any formal collaborative scheduling process—it is difficult to imagine any significant amount of voluntary work being funded privately in the absence of a commitment to use it. Such a collaborative scheduling process would not necessarily contravene the discretion that the government typically insists on having as a sovereign entity. The existence of established test guidelines, GLPs, etc., means that EPA could set objectively verifiable conditions for its commitment, and would only be agreeing to use the study results if the study met those criteria.

Even the chemical industry now agrees that if Congress overhauls the Toxic Substance Control Act (TSCA), industry should have the burden of providing information to show that its chemicals are safe (see ACC). Inherent in such a framework would be industry conducting studies that meet predetermined requirements and submitting them to EPA for its consideration. EPA should be willing to adopt this framework for its IRIS database now.

D. Mechanisms to Allow Agencies to Act Quickly and Yet Retain Flexibility to Change

Notwithstanding the foregoing proposals to coordinate private and public entities on policy-relevant scientific matters, conflicts between the state of the science and the desire for agency action are sure to persist. This last section of the chapter briefly surveys a range of other, more traditionally administrative options for trying to minimize the frequency and, even more important, duration of those conflicts.

1. Don't Insist on Redoing All Values for a Substance Any Time One is Revised

The IRIS database can contain up to four values for a substance, a Reference Concentration (RfC) for noncancer inhalation toxicity, a Reference Dose (RfD) for noncancer ingestion toxicity, and oral slope factors/ inhalation unit risks for carcinogenicity. EPA's current policy is not to revise any one of these singly, but to revise all values for a chemical if it revises any of them. EPA says it does so "for both administrative efficiency and to ensure that we evaluate the breadth of available science" (see, e.g., EPA 2003). It is unclear why the latter requires all existing values to be revised at once or not at all. And given the challenges inherent in setting these values, and the likelihood that the science relevant to any one of them may not be "ready" (in reality or in EPA's view), the practice of tying them all together, while possibly more efficient, is certainly less effective, as it results in freezing the unnecessarily tied values for years longer than need be.

2. Reconsider the Value of Trying to Produce Definitive Pronouncements

As the case studies showed, assessments and reassessments of controversial, well-studied chemicals regularly take decades. If anything, the length of the delays is a negative feedback loop: as it becomes increasingly clear that the issue will be frozen for ten to twenty years, all concerned only heighten their efforts to ensure that the final outcome reflects their perspective—usually producing conflicting desires to speed things up and slow them down. In the meantime, agencies feel growing pressure to take regulatory or policy actions on the chemical. As a result, we see agencies setting values on an ad hoc basis rather than using the "definitive" value that they are supposed to use—for example, the EPA preliminary remediation goals for dioxins.

This dynamic raises the question whether it might actually be superior for agencies to revise health values regularly, if not continuously. Under this approach, when an EPA program office needed to set a value for a chemical, it would take the existing IRIS value, determine what new science had been issued since that value was set, and update it. This is in fact EPA's stated policy regarding IRIS,[16] and the Agency has at times even applied it.[17] That value would then be noted in IRIS. The next time that office or another needed to address the same chemical, it would take the most recently set value and update it. This approach thus would not require starting all over each time, but would rather build on existing work. The recommendations of the NAS in its formaldehyde report that EPA boil its analysis down into standardized evidence tables (NRC 2011, 112–13) would facilitate that process—updated analyses would simply incorporate new work into existing tables and analysis.

Conversely, where an agency is already on a schedule to produce an assessment—particularly a legally enforceable schedule—it would be most efficient for the agency not to revise that value on an interim basis. Thus, it is perverse that EPA has settled a legal challenge to its 2008 ozone National Ambient Air Quality Standard by agreeing to "reconsider" that value in the 2010–2011 timeframe, even though it is statutorily obligated to revise it again in 2013. Even more perverse, and contrary to the proposal made immediately above, the new reconsideration is being limited to the same database used to set the 2008 value: studies published up through 2005, and thus six or more years old. In the crowning irony, EPA is reconsidering the 2008 value using old studies even as it has already begun reviewing the current literature for the 2013 value. Why not consider the new studies now? Why not seek a Congressional exemption from the five-year deadline, or argue that the agency will have met that deadline early? The current approach seems the height of inefficiency and exactly the wrong way to use the newest available science.

3. If a Value is Likely to Change, Sunset Rules Based on It

At times, agencies will conclude that the degree or extent of potential health or environmental risk or other exigencies (including political considerations) simply require action even though there is substantial controversy over whether the science underlying the decision is "ready." As has been seen, agencies in such cases are usually tempted to rush to judgment and claim that the science is readier than it is. Rather than engage in this science charade, however, agencies would do better to concede that the science, while minimally adequate to meet legal standards, could stand to be improved, and then to time-limit their action so that it will expire on a date by which better science is anticipated to have become available. For example, if a major epidemiological study should be completed in a year that might enable a proposed rule to be issued a year later, and a final rule a year after that, then the agency should issue its initial rule with a three-year expiration date. That sunset provision should lessen opposition to the rule deriving from fear that it will live on eternally, based on the soon-to-be-outdated science available when it was issued. The same approach could be applied even if the currently available science is good, if better science was in the works—in fact, the sunset could be the quid pro quo to prompt regulated entities to commit to doing the new scientific work.

4. Experiment with Adaptive Management

Probably the single most useful administrative process for regulating where the science is still evolving is "adaptive management," sometimes summarized as "learning by doing" (Doremus 2007, 550). Simplistically, the conventional regulatory approach seeks to fully understand a scientific issue before acting, but then devotes little effort to evaluating *ex post* whether that understanding was correct, and is biased against modifying the chosen actions in light of their actual effectiveness. By contrast, adaptive management begins by acknowledging that the issue is not fully understood. It then applies an iterative approach of "collecting information, establishing measurements of success, monitoring outcomes, using new information to adjust existing approaches, and [being willing] to change" (Ruhl & Fischman 2011, 429). Ideally, adaptive management applies "an experimentalist research element, in which management actions deliberately probe for information to evaluate testable hypotheses . . . , such as evaluating the effects a chosen habitat management action and its alternatives might have on invasive species by running small-scale test plot experiments" (429). As the previous quote indicates, adaptive management has been most widely used in large-scale ecosystem or resource management projects, and the NAS has issued multiple reports regarding its use in water resources projects (National Academies 2011). Its utility in the area of chemical risk management appears to be much narrower, although a cynic might argue that it is the approach we apply now, particularly to existing chemicals. Also, the actual application of adaptive management often falls short of the theoretical construct described above (Ruhl & Fischman 2010, 427). Nonetheless, as an essentially empirical, experimental approach, adaptive management seems to be obviously superior to the current approach and should be encouraged.

IV. CONCLUSION

Sometimes the science needed to support a regulatory decision is not sufficient at the time an agency wants to make the decision; sometimes it was sufficient long before. Stakeholders in the regulatory process can be expected to take either of these positions where it suits their interests. Disappointingly, the federal government also commonly does so, typically as a strategy for acting on its preferred scientific basis. A variety of criteria, frameworks, and administrative processes exist that could discourage that practice and result in better coordination between the regulatory and scientific timetables—and would save government resources. Use of these tools would, however, require acceptance of greater coordination between government and the private producers of most regulatory science.

NOTES

1. Principal, Conrad Law & Policy Counsel, Washington, DC (www.conradcounsel.com). BA Haverford College, JD George Washington University Law School.

2. Perhaps the most appalling example is the over-400 "permissible exposure limits" or PELs for occupational exposure to air contaminants set by the Occupational Safety & Health Administration (OSHA), most of which were issued in 1971, based on assessments completed in the 1940s–1960s. *See* 29 C.F.R. § 190.1000; AFL-CIO v. OSHA, 965 F.2d 962, 968 nn. 4–5 (11th Cir. 1992). Substantive, procedural, and budgetary restrictions have prevented OSHA from successfully updating more than a handful of these in the intervening forty years.

3. Ioannidis (2005) presents a compelling Bayesian case for this point in 2005. This problem is described less technically in Lehrer (2010, 52).

4. The author apologizes to readers too young and insufficiently retro to know what an Etch-a-Sketch is.

5. In a speech to the assembled U.S. attorneys, then-Attorney General Robert Jackson said: "Your positions are of such independence and importance that while you are being diligent, strict, and vigorous in law enforcement you can also afford to be just. Although the government technically loses its case, it has really won if justice has been done." Jackson (1940, 18).

6. This phenomenon is nicely elucidated in Sarewitz (2004).

7. Id. at 44. Unfortunately, the collateral damage was done; EPA's Air Office had previously proposed to rely on the BBDR model to assess residual risks remaining from hazardous air pollutant emissions from furniture manufacturing, but on the basis of the IRIS office's "extreme" manipulations decided not to. *See* EPA (2010). Thus, the lagging nature of the IRIS process succeeded in reversing adoption of new thinking by an EPA program office.

8. See Cruzan (2011).

9. The SAB member said that EPA's assessment required a "reality check," given "risk values that are pretty high." SAB (2010).

10. The question of what percent of cancer incidence might be accounted for by human contamination of the environment has been estimated by several analysts as between 2–3 percent. See Gough (1990) and Doll and Peto (1981).

11. Id. at 155; see also comments of the panel chair id. at 117 ("[G]iven that there is a fair amount of discomfort both from the public comments and from several members of the SAB around that . . . narrow focus, [the SAB should offer to] do an integrated review . . . , in fact, do what I think everyone really had hoped we would do from the get-go.").

12. www.niehs.nih.gov/news/releases/2009/bisphenol-research.cfm.

13. While the following cases primarily illustrate rushing to judgment, several of the "stalling for time" examples discussed above do as well. For example, much of the public criticism of EPA's 2010 draft inorganic arsenic assessment focused on NCEA's unwillingness to include hundreds of studies published after the 2007 cutoff, including several accepted for or close to publication in 2011 that bear directly on whether a nonlinear cancer slope factor may be appropriate, the key issue highlighted by the 2005 SAB review.

14. cfpub.epa.gov/ncea/iristrac/index.cfm?fuseaction=viewChemical.showChemical&sw_id=1107.

15. cfpub.epa.gov/ncea/iristrac/index.cfm?fuseaction=viewChemical.showChemical&sw_id=1107.

16. Id. ("It is Agency policy that the IRIS represents a starting point for risk assessments, however, it is not given conclusive weight in the context of rulemaking. If an outside party questions information presented in the IRIS, we will consider all credible and relevant information before us in the course of making our decision.").

17. Id. (setting an RfC for methyl ethyl ketone based on different uncertainty factors than those used to set the IRIS RfC); EPA (1999) ("agree[ing] that the IRIS risk estimate is not an appropriate basis from which to extrapolate human risk" and using the draft IRIS value then in progress). As discussed in connection with note xxi supra, EPA's Air Office had proposed to base its estimate of formaldehyde risk on use of a state of the art biologically-based dose response model, rather than the IRIS value, but dropped that approach based on manipulations of the model by NCEA that the NAS found to be "extreme" and said "should not have been used."

REFERENCES

AFL-CIO v. OSHA, 965 F.2d 962 (11th Cir. 1992).

AHR Case Study (2010). "Dose-Response Approaches for Nuclear Receptor-Mediated Modes of Action," (report of workshop conducted September 27–29, 2010), www.tera.org/peer/nuclearreceptor/AHR%20Case%20Study%20Materials%20Document.pdf.

American Chemistry Council (2012). "10 Principles for Modernizing TSCA," www.americanchemistry.com/Policy/Chemical-Safety/TSCA/10-Principles-for-Modernizing-TSCA.pdf [ACC].

American Chemistry Council (2010). "Scientists Say Flawed EPA Dioxin Assessment Ignores National Academy of Sciences, Exaggerates Risk," October 28, americanchemistry.com/s_acc/sec_news_article.asp?CID=206&DID=11449 [ACC 2010].

Association of Metropolitan Water Agencies (2011). "Peer Review Panel Urges EPA to Consider Emerging Cr-VI Research," *Monday Morning Briefing*, May 16.

Boobis, A. R. et al. (2006). "IPCS Framework for Analyzing the Relevance of a Cancer Mode of Action for Humans." *Crit. Rev. Toxicol* 38(2): 87–96.

Butterworth, Trevor (2009). *Science Suppressed: How America Became Obsessed with BPA*, stats.org/stories/2009/science_suppressed_BPA_intro_jun12_09.html.

Comments of Aaron Barchowsky et al. (2010). Docket No. EPA- HQ-ORD-2010-0123 (March 25, 2010), www.regulations.gov/#!documentDetail;D=EPA-HQ-ORD-2010-0123-0004.1.

"Comments of the American Chemistry Council's Chlorine Chemistry Division on NCEA's Review of the University of Michigan Dioxin Exposure Study," (December 22, 2009). www.dioxinfacts.org/NCEA_Cover_%20Ltr_Comments.pdf.

Conrad, James W., Jr, ed. (2011). "Environmental Science Deskbook," §§ 4:4.1, 4:1.

Cruzan, George (2011). "Evaluation of Ramazzini Foundation Study of Methanol in Rats," www.epa.gov/quality/informationguidelines/documents/10005-rel2.pdf.

Department of Labor. Occupational Safety and Health Administration (1989). "Air Contaminants Standard," 54 Fed. Reg. 2332 (January 19, 1989) (codified at 29 C.F.R. § 190.1000) [DL OSHA 1989].

Doll, R., and Peto, R. (1981). "The Causes of Cancer: Quantitative Estimates of Avoidable Risks of Cancer in the Unites States Today." *J. of Nat'l Cancer Inst.* 66: 1191–1308.
Doremus, Holly (2007). "Precaution, Science, and Learning While Doing in Natural Resource Management." *Wash. L. Rev.* 82: 547–79.
Doull, John, and Waddell, Peter (2010). "Balancing Policy with Science." *The Washington Times*, June 22, www.washingtontimes.com/news/2010/jun/22/balancing-policy-with-science/.
Environmental Protection Agency (2012). "IRIS—Basic Information," www.epa.gov/iris/intro.htm [EPA, IRIS Basic Information].
Environmental Protection Agency (2012). "IRIS—Frequent Questions," "How does EPA decide which substances to add or update?" www.epa.gov/iris/help_ques.htm#howsub [EPA, IRIS Frequent Questions].
Environmental Protection Agency (2011). "EPA Announces Plan of Action to Address Four IRIS Assessments," April 11, yosemite.epa.gov/opa/admpress.nsf/48f0fa7dd51f9e9885257359003f5342/f88161843dbdd2a98525786f005b734b!OpenDocument [EPA 4/2011].
Environmental Protection Agency (2011). IG Report No. 11-P-0215, "EPA's Endocrine Disruptor Screening Program Should Establish Management Controls to Ensure More Timely Results," May 3, epa.gov/oig/reports/2011/20110503-11-P-0215.pdf [EPA 5/2011].
Environmental Protection Agency (2010). "IRIS Toxicological Review of Hexavalent Chromium," (External Review Draft). EPA/635/R-10/004A, cfpub.epa.gov/ncea/iris_drafts/recordisplay.cfm?deid=221433 [EPA, Hexavalent Chromium External Review Draft, 2010].
Environmental Protection Agency (2010). "IRIS Toxicological Review of Inorganic Arsenic (Cancer)," (External Review Draft). EPA/635/R-10/001, cfpub.epa.gov/ncea/iris_drafts/recordisplay.cfm?deid=219111 [EPA, Inorganic Arsenic External Review Draft, 2010].
Environmental Protection Agency (2010). "National Emission Standards for Shipbuilding and Ship Repair (Surface Coating); National Emission Standards for Wood Furniture Manufacturing Operations," 75 Fed. Reg. 80228 (December 21, 2010) [EPA 12/2010].
Environmental Protection Agency (2010). "EPA Places Four IRIS Assessments on Hold Pending Review," June 15, yosemite.epa.gov/opa/admpress.nsf/0/b64d44f06a56d5b285257742007c5002?OpenDocument [EPA 6/2010].
Environmental Protection Agency (2009). "Proposed Consent Decree, Clean Air Act Citizen Suit," 74 Fed. Reg. 56612 (November 2, 2009) [EPA 11/2009].
Environmental Protection Agency (2009). "EPA Science Advisory Board Staff Office; Request for Nominations of Experts for the SAB Methanol Review Panel," 74 Fed. Reg. 53740 (October 20, 2009) [EPA 10/2009].
Environmental Protection Agency (2003). "List of Hazardous Air Pollutants, Petition Process, Lesser Quantity Designations, Source Category List," 68 Fed. Reg. 32614 (May 30, 2003) [EPA 2003].
Environmental Protection Agency (1999). "National Air Toxics Program: The Integrated Urban Strategy," 64 Fed. Reg. 38717-38718 (July 19, 1999) [EPA 1999].
Environmental Protection Agency. Science Advisory Board (2011). "SAB Review of EPA's Reanalysis of Key Issues Related to Dioxin Toxicity and Response to NAS Comments," May 4 Draft, yosemite.epa.gov/sab/sabproduct.nsf/ea5d9a9b55cc319285256cbd005a472e/bbb706d0201bc5e485257888006f3f11/$FILE/Dioxin%20for%20Qual%20Review%205-4-11.pdf [EPA SAB 5/2011].
Environmental Protection Agency. Science Advisory Board (2011). "Review Comments on EPA's Responsiveness to SAB 2007 Recommendations for the Revision of Cancer Assessment of Inorganic Arsenic," March 1, yosemite.epa.gov/sab/sabproduct.nsf/0/9FCEE4E20ABD6EB48525784600791AC2/$File/EPA-SAB-11-003-unsigned.pdf [EPA SAB 3/2011].

Environmental Protection Agency. Science Advisory Board (2010). "Quality Review Teleconference." November 22 transcript [EPA SAB 11/2010].
Environmental Protection Agency. Science Advisory Board (2007). "Advisory on EPA's Assessments of Carcinogenic Effects of Organic and Inorganic Arsenic," (EPA-SAB-07-008), June 28, yosemite.epa.gov/sab/sabproduct.nsf/02ad90b136fc21ef85256eba00436459/EADABBF40DED2A0885257308006741EF/$File/sab-07-008.pdf [EPA SAB 2007].
Food & Drug Administration (2010). "Update on Bisphenol A for Use in Food Contact Applications," www.fda.gov/NewsEvents/PublicHealthFocus/ucm197739.htm [FDA 2010].
Food & Drug Administration (2008). "Draft Assessment of Bisphenol A for Food Contact Applications," www.fda.gov/ohrms/dockets/ac/08/briefing/2008-0038b1_01_02_FDA%20BPA%20Draft%20Assessment.pdf [FDA 2008].
Food & Drug Administration (2007). "FDA Statement on European Aspartame Study," April 20, www.fda.gov/Food/FoodIngredientsPackaging/FoodAdditives/ucm208580.htm [FDA 2007].
Goodman, J. E. et al. (2006). "An Update Weight of the Evidence Evaluation of Reproductive and Developmental Effects of Low Doses." *Crit. Rev. Toxicol.* 36(5): 387–457.
Gough, Michael (1990). "How Much Cancer Can EPA Regulate Away?" *Risk Analysis* 10(1): 1–6.
Gray, G. et al. (2001). "The Annapolis Accords on the Use of Toxicology in Risk Assessment and Decision-Making," www.accp1.org/pdf/TheUseofToxicology1.pdf.
Guzelian, P. S. et al. (2005). "Evidence-Based Toxicology: A Comprehensive Framework for Causation." *Hum. Exp. Toxicol.* 24(4): 161–201.
Hill, Austin Bradford (1965). "The Environment and Disease: Association or Causation?" *Proc. R. Soc. Med.* 58: 295–300.
Hites, Ronald A. (2011). "Dioxins: An Overview and History." *Environ. Sci. Technol.* 45(1): 16–20.
Hoffmann, S., and Hartung, T. (2006). "Toward an Evidence Based Toxicology." *Hum. Exp. Toxicol.* 25(9): 497–513.
Hogue, Cheryl (2010). "More Dioxin Delays." *Chem. & Engr. News* 88:30 (paraphrasing Harvey Clewell), pubs.acs.org/cen/government/88/8846gov1.html.
International Epidemiological Association (2007). "Good Epidemiological Practice (GEP): IEA Guidelines for Proper Conduct in Epidemiologic Research," www.ieaweb.org/index.php?option=com_content&view=article&id=15&Itemid=43.
Ioannidis, John P. A. (2005). "Why Most Published Research Findings Are False." *PLoS Med.* 2(8): 0696–0701, www.plosmedicine.org/article/info:doi/10.1371/journal.pmed.0020124.
Jackson, Robert H. (1940). "The Federal Prosecutor." 24 *J. Am. Jud. Soc'y* 18.
Jasanoff, Sheila (1990). *The Fifth Branch.* Cambridge, MA: Harvard University Press.
Klimisch, H-J. et al. (1996). "A systematic approach for evaluating the quality of experimental toxicological and ecotoxicological data." *Regul. Toxicol. Pharmicol.* 24: 1–5.
Kovski, A. (2010). "EPA Review of Atrazine Health Risks Prompts Questions from Advisers, Others." *Daily Env't Rept,* October 14, A-12.
Lehrer, Jonah (2010). "The Truth Wears Off." *The New Yorker,* December 13, 52.
"Letter from Thomas Zoeller et al. to Margaret Hamburg," (September 21, 2009), www.weaversway.coop/blog/Letter%20to%20Commissioner%20Hamburg.pdf.
Michaels, David M. (2008). *Doubt is Their Product.* New York: Oxford University Press.
Mukherjee, Siddhartha (2010). *The Emperor of All Maladies.* New York: Scribner.
National Academies. Water Information Center (2011). "Adaptive Management," water.nationalacademies.org/adapt2.php [NA 2011].
National Institute of Environmental Health Sciences. National Institute of Health (2011). "Questions and Answers about Bisphenol A," (September 8, 2011), www.niehs.nih.gov/news/media/questions/sya-bpa.cfm [NIEHS 2011].

National Institute for Environmental Health Sciences (2010). "Dose-Response Approaches for Nuclear Receptor-Mediated Modes of Action," (report of workshop conducted September 27–29, 2010), www.tera.org/peer/nuclearreceptor/AHR%20Case%20Study%20Materials%20Document.pdf [NIEHS 2010].

National Research Council. 2011. "Review of the Environmental Protection Agency's Draft IRIS Assessment of Formaldehyde (NAS Report)," www.nap.edu/catalog.php?record_id=13142 [NRC, NAS Report 2011].

National Research Council (2011). "Science for EPA's Future, Project Scope," www8.nationalacademies.org/cp/projectview.aspx?key=49365 [NRC 2011].

National Research Council (2009). *Science and Decisions* [NRC 2009].

National Research Council (2007). *Toxicity Testing in the 21st Century: A Vision and a Strategy* [NRC 2007].

National Research Council (2006). *Health Risks from Dioxin and Related Compounds: Evaluation of the EPA Reassessment*, dels.nas.edu/Report/Health-Risks-from-Dioxin-Related/11688 [NRC 2006].

National Research Council (1983). *Risk Assessment in the Federal Government: Managing the Process* [NRC 1983].

National Toxicology Program (2005). *11th Report on Carcinogens*, "Introduction," ntp.niehs.nih.gov/ntp/roc/eleventh/intro.pdf [NTP 2005].

National Toxicology Program. NTP Report on Carcinogens Review Process, "Expert Panel Meeting," (last updated October 3, 2011), ntp.niehs.nih.gov/index.cfm?objectid=FA93DE50-F1F6-97

Office of Management and Budget (2005). Final Information Quality Act Bulletin for Peer Review. Federal Register 70: 2664, 2669.

Sharpe, R. M. (2010). "Toxicological Highlight: Is it Time to End Concerns over the Estrogenic Effects of Bisphenol A?" *Toxicological Sciences* 114(1): 1–4.

vom Saal, F. S. et al. (2007). "Chapel Hill Bisphenol A Expert Panel Consensus Statement: Integration of Mechanisms, Effects in Animals and Potential to Impact Human Health at Current Levels of Exposure." *Reproductive Toxicology* 24(2): 131–38.

SEVEN

Improving the Use of Science to Inform Environmental Regulation

Susan E. Dudley and George M. Gray

Scientific information is essential for making sound public policy decisions, particularly those related to health, safety, and the environment. Policy makers depend on assessments of the likely risks associated with different actions (or inaction) in order to make decisions about what actions can most effectively improve public health and the environment. Assessments of the risks posed under different conditions and of the potential changes in risk achievable from different policy options depend not only on pure scientific inputs, but on scientific assumptions and judgments. Scientists will never have complete information to predict outcomes with absolute certainty, so risk assessments embody science policy choices—assumptions and rules of thumb—that guide the use of scientific information in analyses that inform policy in the face of this uncertainty.

Policy makers and the public are often unaware of the influence of these science policy choices or the existence of alternative, scientifically plausible, assessments. Instead, assessments often generate precise-sounding predictions that hide considerable uncertainty about the actual risk. Thus policy decisions that are thought to be based on science are heavily influenced by hidden judgments about how to deal with uncertainty. Furthermore, even if it could be done perfectly, assessment of risk alone is not sufficient for establishing policy to address that risk. A second analytical step, "risk management" relies on risk assessment as well as other information to answer the policy question, "how much risk is too much?" While risk assessment is essential for understanding the positive question of *what is*, or predicting what outcomes might derive under different scenarios, it is less helpful for the normative (policy) decisions regarding what *should be*. (See figure 7.1.) Yet virtually every participant in the regulatory development process participates in what Wagner (1995) calls the "science charade" pretending that policy can be made purely on scientific grounds, and not revealing the real reasons for choosing one policy over another.

As the Bipartisan Policy Center's (2009, 10) report, "Improving the Use of Science in Regulatory Policy" observed, "a tendency to frame regulatory issues as debates solely about science, regardless of the actual subject in dispute, is at the root of the stalemate and acrimony all too present in the regulatory system today."

Institutional arrangements in the regulatory development process tend to aggravate these problems, perpetuating the science charade, insulating experts involved in a particular rulemaking from dissenting views, reinforcing preconceptions and biases, and leading to regulatory policy decisions that are not at all transparent.

Our goal in this chapter is to explore the motivations of the participants in the development of science-based regulation (from legislators to the courts) in order to understand their role in perpetuating the science charade. We then use these observations to identify procedural and institutional approaches that may serve to counter the perverse incentives inherent in today's institutions and improve not only how science informs regulatory policy but scientific knowledge itself.

This chapter first provides background on the role of science and science policy in regulatory decisions, and suggests that current practices lack transparency, resulting in false precision and a blurring of science and policy. It then reviews the regulatory development process and examines the motivations of each participant to "frame regulatory issues as debates solely about science" (Bipartisan Policy Center 2009, 11), and perpetuate the science charade. It concludes with recommendations for

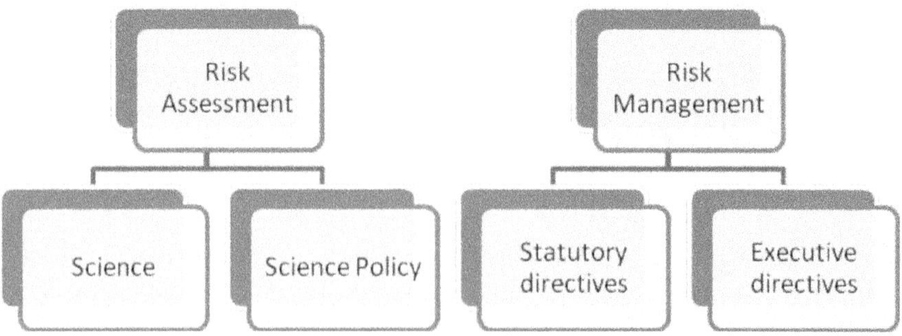

Figure 7.1. Risk Assessment and Risk Management

addressing the problems identified here. Our focus is on regulations issued by the Environmental Protection Agency, though we sometimes refer to other agencies for comparison.

SCIENCE, SCIENCE POLICY, AND POLICY

The use of science to inform important social decisions in health, environment, and energy policy is critical to sound and legitimate policies. The credibility of a variety of important decisions—from regulations on pesticides to the use of medical devices—is predicated on the appropriate use of scientific information. Yet there is general agreement that the regulatory science policy system is broken (Bipartisan Policy Center 2009). Experts have raised questions about how science is analyzed and presented in fields as diverse as climate change (Reilly, Stone, Forest, Webster, Jacoby, and Prinn 2001), chemical risks (National Research Council 2010), and airline safety (Hough 2010). In most cases, a key concern is that risk assessments do not acknowledge the range of potential outcomes that reflect the uncertainty in underlying scientific data, theories, and models. Scholars and expert panels have offered numerous suggestions for better characterization of uncertainty to increase the transparency of risk assessment (e.g., National Research Council, 1983; National Research Council 1994; National Research Council 2009) but, perhaps due to the incentives discussed below, some of EPA's risk assessments continue to be criticized for continuing to use approaches that obfuscate important uncertainties and conceal legitimate scientific ambiguity (National Research Council 2006; National Research Council 2010; National Research Council 2011).

Some Definitions

Before addressing opportunities to improve the characterization of science to inform policy, some definitions and explanations are necessary. We make a distinction between science, science policy, and policy (see figure 7.2). As noted above, *policy* reflects normative values regarding what should be, and involves a weighing of various factors, including information generated through science and science policy judgments. *Science* is the act of collecting and analyzing data by conducting measurements, modeling processes, and carrying out experiments. The process of scientific inquiry revels in debate, discussion and discourse. Analyses and results are usually subject to peer review and shared through scholarly publications, reports, and professional meetings.

Science policy is concerned with setting up rules for how scientific information will be used in analyses to inform social decisions. In many cases, this occurs through the regulatory risk assessment process.[1] The distinction between science and science policy in risk assessment and

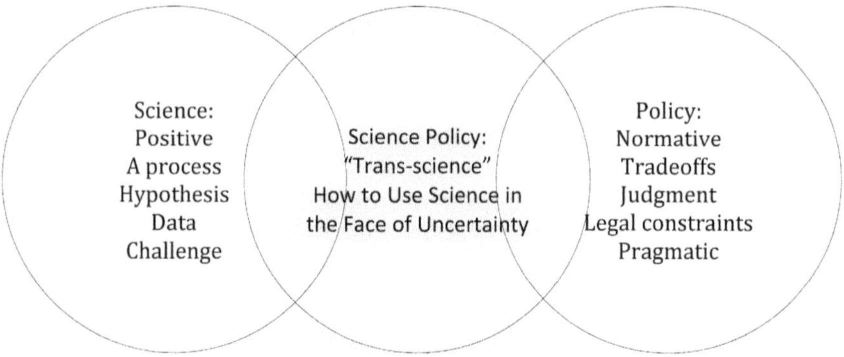

Figure 7.2. Science, Science Policy, and Policy

regulatory decision-making is frequently misunderstood or even deliberately obfuscated (Graham 1996).

In risk assessment, science policy choices include which science is considered, how individual studies are weighed and combined, when competing theories are considered appropriately supported for inclusion, which models to use, and in general, what to do in the face of scientific uncertainty. Additionally, science policy guides the way in which risks are characterized and communicated. These choices, which can have a great influence on the results of a risk assessment and its role in decision making, are ultimately not scientific choices but choices guided by a particular philosophy about how to deal with uncertainty. The U.S. EPA has long recognized the need for science policy choices; its 2001 *Risk Characterization Handbook* reminds agency risk assessors that:

> Importantly, remember that risk characterization is not just about science. It makes clear that science doesn't tell us certain things and that science policy choices must be made. (USEPA 2001, 11)

At the EPA, science policy for risk assessment is codified in guidance. These documents, such as the *Guidelines for Carcinogen Risk Assessment* (USEPA 2005) or *Guidelines for Developmental Toxicity Risk Assessment* (USEPA 1999) provide agency risk assessors and others doing assessments for or to be presented to the agency with principles to inform identification, interpretation, and analysis of scientific data. Because these choices and assumptions are about how to use science, they require an underlying philosophy to guide decisions. EPA has stated its philosophy as follows:

> First, there are some basic, fundamental policy positions that frame the risk assessment process to ensure that the risk assessments produced are appropriate for a particular decision. For example, since EPA is a health and environmental protective agency, EPA's policy is that risk

> assessments should not knowingly underestimate or grossly overestimate risks. This policy position prompts risk assessments to take a more "protective" stance given the underlying uncertainty with the risk estimates generated. (USEPA 2004, 13–14)

DEFAULT ASSUMPTIONS AND THE TYRANNY OF THE SINGLE NUMBER

The role of science policy and the appropriateness of conservatism (the "protective" stance) in risk assessment are subject to considerable discussion and debate.[2] Much of this has focused on the use of "default" choices or assumptions for key risk assessment judgments in the face of uncertainty. These judgments include, for example, which test species to rely upon, which endpoint to consider,[3] how to extrapolate response from high dose animal studies to the low doses encountered by people (often a 1000 to 10,000 fold difference), or how to estimate exposure to an agent (Gold, Stern, Slone, Brown, and Manley 1997, 117). Often agencies adopt default assumptions, with a goal of being "protective," that are deliberately erring on the side of overstating, rather than understating risk. For example, if a rat assay indicates a higher response to exposure to a particular chemical than a mouse assay, the rat data would be the default used in agency risk assessments for that chemical. The most recent EPA risk assessment guidance has stated a preference for using the choice with the greatest scientific backing in place of defaults (U.S. EPA 2005). Only if the agency judges none of the competing scientific views is sufficiently supported should a default be invoked (although the National Academies of Science has questioned how well this policy is implemented) (National Research Council, 2009).

A key point, however, is that by having a science policy that embraces a *single* choice from scientifically plausible alternatives, current risk assessment procedures hide the quantitative implications of alternative choices from risk managers and the public (Ohanian et al., 1997). In a recent review of EPA's reassessment of the risks of dioxin, a National Academies of Science National Research Council panel stated:

> The EPA [Dioxin] Reassessment does not explain why EPA chose to rely on a single site (liver) from one sex (female) in one species (rat), as measured in a single study (Kociba et al. 1978). Consideration of other data sets that were available to EPA would have yielded a substantially wider range of potency estimates. (National Research Council 2006, 90)

The panel followed by pointing out that this single answer approach to assessment might be misleading:

> While EPA qualitatively notes many of the factors contributing to this estimate's uncertainty, the [Dioxin] Reassessment does not adequately discuss how these factors contribute quantitatively to the underlying

> uncertainty. By omitting the quantitative implications of these factors, the Reassessment understates the uncertainty inherent in these estimates and overstates the consistency of the data and risk estimates across all studies. (National Research Council 2006, 91)

In some cases, EPA has made considerable effort to quantify important sources of uncertainty (e.g., U.S. EPA, 2006) in analyses, although these have been faulted as well (Fraas 2010). Development of guidance for more advanced uncertainty analysis may help in the future (U.S. EPA, 2009b).

Our point is not to debate the underlying philosophy of risk science policy, but to identify how the single estimate of risk developed with current science policy may be undermining the scientific credibility of agency assessments and hindering transparency about the underlying science for decision makers and the public. Our goal is ultimately to suggest solutions that will help agencies enhance risk assessment practice and its utility.

There is clearly a disconnect between agency science policy approaches and the way the broader scientific community believes risks should be analyzed. Concerns relate to how EPA evaluates studies[4] (ultimately to choose a single study for its assessment) and how well EPA presents true scientific disagreement about the appropriate approach to estimating dose-response relationships[5] (again, by endorsing a single approach and single outcome). If EPA's risk assessments do not reflect the evolution of scientific knowledge, both the credibility of the assessments and the development of new scientific information are threatened. If advances in understanding are not reflected in assessments, support for that research, in both the public and private sectors, may diminish.

Ultimately, some form of science policy is necessary for consistency in assessments to support regulatory actions. However, current science policy for risk assessment leads to false precision in the presentation of scientific information and thereby threatens the scientific credibility of the process, hides rather than makes transparent the uncertainty in assessments of risk, and enables the "science charade" by putting key policy choices in the hands of risk assessors and allowing policy makers to avoid making hard decisions. Understanding what motivates participants in the rulemaking process may help identify solutions to these problems.

PARTICIPANTS IN THE RULEMAKING PROCESS, THEIR MOTIVES AND BEHAVIOR

In this section, we review the process for developing regulations to explore what might motivate different parties to conceal uncertainty and blur the roles of policy and science.

The development of regulation in the United States involves several steps and numerous parties. First, Congress must pass and the president must sign legislation authorizing regulation. Modern legislation generally expresses broad goals and objectives, but leaves the details of implementation to executive branch agencies, such as the EPA (Schoenbrod 1995). Regulatory agencies then develop draft proposed regulations consistent with the language in the enabling legislation and according to procedures mandated by both Congress and the president (Lubbers 2006). In particular, the Administrative Procedure Act requires regulatory agencies to seek public comment on proposed regulations, and to base final regulations on information in the rulemaking record (Dudley 2010). This notice-and-comment process guarantees interested parties (those affected by potential regulation and non-governmental organizations) an opportunity to present views and information on proposed regulations. Additionally, modern presidents have required agencies to conduct regulatory impact analyses of economically significant regulations, and to subject them to interagency review through the Office of Information and Regulatory Affairs (OIRA) in the Office of Management and Budget.[6] Congress has an opportunity to issue a joint resolution of disapproval after a final regulation has been published,[7] and regulations are also subject to judicial review (where affected parties can sue to have regulations overturned by the courts). Throughout the rule development process and beyond, the media also track and report on regulations and any controversies that may arise.

The behavior of each party in the regulatory development process is influenced by these institutional frameworks and constraints, and the incentives they provide. Understanding these institutional incentives can help explain what motivates participants at each stage of regulatory development to engage in the science charade, and will be essential to identifying solutions to the problems identified above.

Legislators

Congress is a political body, and members are generally understood to be motivated by the desire to be reelected. In the face of a crisis, elected officials have strong incentives to "do something." Legislative action can range from conducting hearings, investigations, and interventions to motivate executive branch officials to pursue particular actions (often through regulation under existing statutory authority), or using legislation and appropriations to authorize new regulatory and enforcement activity. Whether the regulatory action they demand ultimately produces the desired outcomes is less important from a legislator's perspective than the appearance of action, in large part because those effects are not immediately apparent, but also because action simply appears more constructive than inaction.

In many cases, harsh rhetoric, combined with admirable-sounding legislative mandates, allows members to take credit for responding to crises (Schoenbrod 1995). What may be missing, however, is a feedback mechanism. Have the actions undertaken pursuant to legislative mandates had the desired effects? Were there unintended consequences?

Complexity and nuance are not easily conveyed to voters, and the most effective political messages tend to be simple ones that reinforce (or cater to) core values and beliefs. Thus, black-and-white goals such as protecting public health or saving children are more easily communicated than goals that acknowledge uncertainty in our scientific understanding of risk, or tradeoffs between competing objectives, such as risk reduction and economic growth.

Elected officials generally do not have the scientific or technical expertise[8] to identify how to achieve desired goals, so statues will often direct regulatory agencies to rely on "science" to develop implementing regulations. They fail to recognize the limits of science to define problems and identify cause and effect, or to appreciate the uncertainty inherent in scientific assessments or the inevitable mix of science, science policy, and policy choices involved in regulatory decisions.

While the United States has not formally adopted the precautionary principle as a policy, the science policy judgments buried in risk assessments take a "'protective' stance" as described above (Brock, Rodricks, Rulis, Dellarco, Gray, and Lane 2003). Furthermore, many statutes implicitly adopt a precautionary approach. For example, the Endangered Species Act, sections of the Clean Air Act, the Food, Drug and Cosmetics Act and others call for the elimination of unreasonable risk, without recognition of tradeoffs or unintended consequences.[9] Such statutes generally do not call for zero risk, which is obviously unachievable, but instead use language such as "reasonable risk" and leave to "science" the job of defining what is reasonable.

Such legislation rarely recognizes that determinations of what is "reasonable" involve tradeoffs. Though people routinely make such tradeoffs in daily life (whether to eat a dessert instead of extra vegetables, to buy the larger, safer car or one that is more fuel efficient or less expensive), legislators run the risk of appearing callous if they publicly consider them in establishing policy. It is easier to portray important policy decisions—who to protect, how completely to protect, how to determine "requisite"—as determinations of science, rather than recognize the uncertainties and tradeoffs involved.

As long as voters uncritically accept promises of regulatory benefits without tradeoffs, the incentives for members of Congress are clear. Through vaguely written statutory language with noble goals, members can claim credit from their constituents for demanding safer mines, greater homeland security, or air quality that is requisite to protect public health and welfare, while deferring the details to executive branch regulators. As an added advantage, they can reserve the option of blaming any undesirable outcomes on how the goals were implemented by agencies.[10]

Policy Officials

Like their elected counterparts in Congress, policy officials (used here to mean political appointees in the executive branch) face pressure to act in response to a crisis. The appearance of being responsive is often more important than whether the response is effective. Unlike legislators, however, executive branch policy officials are responsible for implementing the broad policy goals expressed in statutory mandates, and must provide specifics on how they will be achieved.

Like all individuals, policy officials have incentives to protect their reputation and gain the respect of others, and this may manifest itself in several ways. First, they may engage in a type of risk-aversion that Wilson refers to as avoidance of "learned vulnerabilities" (Wilson 1989, 192). In regulating unknown risks, regulators (career staff as well as policy officials) face two possible types of errors: (1) permitting the production of something that turns out to have unanticipated risks, or (2) not permitting the introduction of a safe and valuable product. Huber suggests they are much more vulnerable if they make the first type of error (Huber 1983). If a policy official approves a product that later turns out to have adverse effects, she faces the risk of being dragged before Congress and pilloried by the press. On the other hand, since the potential benefits of a new product or technology are not widely known, the risks of disapproval (or delays in approval) are largely invisible, so the consequences of delay are less severe.

A second factor influencing policy officials' behavior relates to their lack of expertise in risk assessment or science policy, which makes them unlikely to fully appreciate the true degree of uncertainty in risk assessments (Felter and Dourson 1998), and unwilling to question underlying assumptions or data.

In her confirmation hearing in January 2009, EPA Administrator Lisa Jackson promised to defer to "the expert judgment of the agency's career scientists and independent advisors."

> If I am confirmed, I will administer with science as my guide. I understand that the laws leave room for policy-makers to make policy judgments. But if I am confirmed, political appointees will not compromise the integrity of EPA's technical experts to advance particular regulatory outcomes. (Jackson 2009)

This belies the complexity and uncertainty involved in "advancing particular regulatory outcomes" and the range of plausible outcomes depending on which data, models, and assumptions are considered. This statement seems to presume a monolithic "scientific answer" without recognizing that there may be legitimate disagreements about agency science policy choices.

Third, while policy officials are not motivated by the desire to be reelected, their reputations among certain groups do matter. Since they are likely to hold their position for a relatively short time, two-and-a-half years on average, (Wilson 1989, 322) they may be looking to their next career move, and how future employers will view their experience and actions. They may align themselves with "client" interests (outside the government), the president or other policy officials, congressional allies, or career staff of agencies.[11]

Policy officials who view the agency as a stepping stone for their future careers are likely interested in gaining the support of (or at least not alienating) interest groups, which could be regulated entities or non-profits interested in particular areas. They may not have an incentive to question the scientific basis behind staff recommendations because their future effectiveness will depend on their relationship with the regulatory agency. Both because career staff will remain involved in policy long after the policy official has left, and because of the deference career staff (scientists in particular) are given in the media and in Congress, policy officials have little incentive to question their recommendations, even when they are aware of the uncertainties behind risk assessments.

Pressure from other policy officials (particularly those with staffs who also have scientific expertise and face countervailing statutory authority and interests) and the president may lead policy officials in regulatory agencies to question single point estimates and decision criteria. However, presidential directives to coordinate with other executive branch agencies and examine tradeoffs are often not effective at challenging decisions that purport to be based solely on science because executive directives are generally subordinate to statutory directives.[12] Furthermore, agency policy officials face the same incentives their elected counterparts do to "do something" in the face of public concern, and the same short time horizon that places greater value on easily communicated short-run fixes than long-run solutions that recognize tradeoffs and unintended consequences.

Thus, policy officials regularly defend their decisions as being based on "the best science available," rather than admitting the degree of uncertainty imbedded in risk assessments or the tradeoffs involved in the risk management decision. When stakeholder questions arise, they may be portrayed as outliers or biased views rather than legitimate disagreements about the agency's science policy choices. Particularly when statutes do not explicitly require regulators to consider tradeoffs in setting standards, admitting to considering such tradeoffs can lead to litigation that overturns a regulation (Wagner 1995, 1666).

Civil Servants

Congress and policy officials rely heavily on career professionals in the civil service for the analysis that underpins regulatory decisions, and grant them significant deference on matters characterized as science.

This deference poses several potential problems. First, as discussed above, it does not recognize the extent of the uncertainty inherent in assessing risks, and the science policy decisions that must be made at each step. Second, it presumes that science alone can answer policy questions. Even if our assessments of risk were completely certain, understanding risk alone, without knowing the tradeoffs involved in reducing those risks (tradeoffs that include increasing other risks), is not usually determinative regarding how much risk is acceptable.

Third, this deference does not recognize that career scientists, like all people, have policy preferences and biases. They are more likely to have opportunities to impose their values when policy decisions are characterized as technical science questions.[13] In conducting analyses and applying science policy, controlling the assumptions means controlling the outcome. In characterizing risks, civil servants may have incentives to understate the uncertainty inherent in risk assessments and to produce quantitative predictions that leave policy officials little room to deviate from staff-presented policy options. Policy officials who do not heed precise-sounding warnings of dire outcomes if certain regulatory pathways are not followed run the risk being accused of, in Jackson's words "compromis[ing] the integrity of EPA's technical experts to advance particular regulatory outcomes."

This does not suggest that career staff have malevolent motives. In the face of science-based statutory mandates, if they acknowledge uncertainty or admit to considering other factors, they risk seeing their regulations overturned in court. Further, agencies would likely find it difficult to defend a shift in practice to produce a central estimate of risk that was significantly lower than previously reported estimates based on "protective" default assumptions. Any attempt to acknowledge scientific uncertainty and present a more accurate risk estimate could be portrayed as "rolling back protections" or make an agency subject to attacks and perhaps lawsuits by parties that relied upon the earlier, more conservative, estimates.

As Justice Stephen Breyer observes in his 1993 book *Breaking the Vicious Circle*, "well-meaning, intelligent regulators, trying to carry out their regulatory tasks sensibly, can nonetheless bring about counterproductive results" (Breyer 1993, 11). Breyer attributes this problem to a combination of public perceptions, congressional actions, and uncertainties inherent in understanding and predicting risks. These external factors exacerbate the problem of "tunnel vision," a phrase he uses to describe how agencies single-mindedly pursue a particular goal to a point that "the regulatory action imposes high costs without achieving significant additional safety benefits."

These incentives are compounded by the organizational structure of federal agencies, which tends to be focused on a single mission (homeland security, safe drugs, clean environment). Agencies are usually organized into smaller units (divisions and branches) in which individuals with narrower missions (such as reducing exposure to an individual substance) work closely together. While cohesive units focused on a shared goal have advantages (Wilson 1989), they are also susceptible to what Janis called "groupthink" (Janis 1972), where group dynamics lead to poor decision making. A broad literature suggests that groups of individuals, when exposed to others with similar views, tend to become more polarized and extreme in their views (Sunstein 2009).

"In a wide variety of experimental contexts, people's opinions have been shown to become more extreme simply because their initial views have been corroborated and because they have been more confident after learning of the shared views of others" (Baron, Brunsman, Hoppe, Kao, Linneweh, and Rogers 1996, 537). The symptoms of groupthink include "close-mindedness, involving a collective effort 'to rationalize' so as to discount warnings or information that might lead to reconsideration, and stereotyped views of enemies, as too evil to warrant efforts at negotiation or 'too weak and stupid to counter' the group's risky choices" (Sunstein 2009, 86). Agency colleagues who work together on a particular policy become more polarized because they do not sufficiently appreciate each other's biases (or recognize that their views may not reflect those of a broader community).

Efforts to bring a broad range of perspectives and considerations to bear on policy questions have met with limited success. Executive requirements for analysis and review are minimized for decisions that are characterized as scientific, and as a result, the role for other agencies, and indeed non-science expertise within an agency, is minimized.

Advisory Boards

One approach to counteracting the problems of tunnel vision and groupthink in agencies is to require peer review of the scientific analysis used to support decisions. Agencies that rely on scientific data have policies regarding peer review.[14] In addition, agencies turn to outside advisory boards to evaluate and opine on the scientific basis for risk assessments and other analyses. The federal government uses over one thousand federal advisory committees to address a wide range of issues, including scientific and technical issues related to health, safety, and the environment (Federal Advisory Committees Database). The Food and Drug Administration relies on forty-nine "committees and panels to obtain independent expert advice on scientific, technical, and policy matters" (U.S. Food and Drug Administration). Some statutes, notably sections of the Clean Air Act, authorize a specific role for outside advisory boards (U.S. EPA 2009).

Given the influence these boards have on important policy decisions, concerns may be raised about conflicts of interest and bias. Advisory panel members may have their research funded by the agency under review or be asked to evaluate an Agency's use of their own research. Moreover, the lines between scientific advice and policy judgment are not always clear. Board members may suffer from the same group dynamic problems discussed above, and have strong incentives to justify opinions as science-based and any disagreement as anti-science or politicized science.

Business Interests

Interested parties (e.g., lobbying organizations and those affected by the rule) often have several opportunities to communicate with regulatory agencies during the development of a proposed rule. At a minimum, the APA requires agencies to provide broad public notice of their intended actions by publishing proposed rules in the *Federal Register* and anyone has an opportunity to submit comments on the proposal and the supporting documentation at that time.

Generally, businesses are concerned with how a regulation will affect their bottom line. This does not mean they will always oppose more stringent regulation, however, as regulations can confer competitive advantage (Stigler 1971). For example, a chemical company might support a ban on a substance for which it has a patent for a substitute. Companies might also not object to a regulation that, while costly, would impose larger costs on competitors.

Though they recognize tradeoffs (and indeed are concerned with the effect of the policy on their long-run viability), business interveners in a developing regulation often perpetuate the science charade by arguing their case, not on economic or practical grounds, but based on science. This focus is sometimes necessary given institutional constraints. For example, under the Clean Air Act, the EPA Administrator cannot consider economic consequences in setting national ambient air quality standards, so only scientific arguments have any prospect of influencing the decision. Business interests also have public relations reasons for focusing on science. They don't want to appear crassly motivated by profits.

Given the significant uncertainty inherent in risk assessment, it is in the interest of potentially affected parties to present alternative analyses and outcomes. Rarely are the differences in science policy judgments exposed in a transparent way, however. Wagner notes that advocates

> . . . rarely, if ever, . . . challenge the agencies' consistent overstatement of the scientific bases for fundamental policy choices. Instead, they become single-mindedly engaged in presenting opposing scientific justifications, demanding outside scientific review, or attacking the competence of the agency's science when it leads to results that run counter to their own unexpressed policy preferences. (Wagner 1995, 1657)

We tend to see a battle of the experts, rather than a serious discussion of what effect different assumptions and science policy judgments have on the resulting policy. When making a case before the court of public opinion, it is easier to say "my expert has better credentials than yours" than to educate the media and public about the uncertainties in agency models, or to explain the tradeoffs involved.

Non-Governmental Organizations (NGOs)

Similar to business interests, NGOs (such as environmental or public health advocacy groups) have opportunities to weigh in on regulatory policies and the scientific justification for them. They tend to be motivated by reputation and membership rather than by profits, but also have strong incentives to focus on science. Similar to members of Congress, NGOs raise their reputations when they successfully lobby for legislation or regulation with high-minded goals (such as reducing asthma-aggravating smog, or protecting the public from the dangers of genetically modified foods). Membership and financial support for their efforts depend on raising concern about the risks of unregulated products and activities and they have little incentive to temper their message with explanations of uncertainty in risk estimates or potential negative consequences of a regulatory action.

So, NGOs also have incentives to engage in the science charade, and enter their experts into the battle over science. As discussed below, the media is generally more sympathetic to the NGO perspective, and their arguments are more compelling to the broader public. They have what Sunstein refers to as a "rhetorical advantage" (Sunstein 2009).

Wagner notes:

> This strategic behavior by interest groups, who are believed by administrators and legislators to oversee the agencies and keep them accountable to the public, has the contrary effect of lending still greater legitimacy to the agencies' science charade. The public, Congress, the courts, and any skeptics remaining inside the agency are left with little doubt that the dispute will ultimately rest with the scientific experts. (Wagner 1995,1661)

Media

The media plays a key role in communicating the policy decisions and debates to the broader public. Since public perception matters to legislators, policy officials, businesses, and NGOs, how the media portrays policy decisions and the science behind them is very important.

Toby Murcott, science correspondent for the BBC, is critical of the "priesthood" of science writers, who merely translate to the masses the words of the "deities" (scientists), without critical evaluation (Murcott 2009). He argues that "genuine public awareness of science [demands] an understanding of how scientific knowledge is crafted." And yet, this is difficult to achieve. Writers generally have a short amount of time to cover a particular story, and simply don't have the resources to understand and communicate the complexity and uncertainty involved in policy decisions. When writing about environmental or public health risks, short sound bites are more likely to capture attention than careful explanations of uncertainties and tradeoffs.

Journalists are likely to fall back on what public relations professionals refer to as the "3-Vs." Engaging stories have a Victim, a Villain, and a Vindicator. For environmental and health regulation, victims are often not identifiable individuals (as risk assessment predicts "statistical" morbidity and mortality) but rather population segments (e.g., children whose IQ may suffer from lead exposure, or patients who suffer side effects from a drug). Villains are often the business that produces products with externalities or side effects. Government regulators can be villains or vindicators (and they would obviously much prefer the vindicator role). They are more likely to be villains if they regulate less than some would prefer or make Huber's first type of error, by approving a product that has unexpected risks, and vindicators if they propose aggressive regulation. NGOs are most often portrayed as vindicators who defend the public interest against business interests.

Individual messengers and messages matter, as well. The views of a scientist are more likely to be treated deferentially than the views of a politician or business representative. With regard to the message, research suggests that people, when presented with different estimates of risk, are likely to believe the worst case scenario, regardless of whether it comes from the villain, victim, or vindicator, rather than to evaluate the merits of different risk estimates (Viscusi 1997). Viscusi found that individuals' responses to divergent risk information revealed "extreme violations of rationality," as individuals place "inordinate weight on the high risk assessment." He notes that "media and advocacy groups often highlight the worst case scenarios, which tend to intensify" observed biases in weighting worst case scenarios more heavily than experts would consider rational.

Courts

After an agency issues a final regulation, parties unhappy with the outcome can sue to have the regulation altered or overturned. Thus the judicial branch often has the last say in interpreting statutory intent and whether the regulation met APA standards.

Wagner's review of case law suggests "that the courts are exacerbating, rather than discouraging, the agencies' misidentification of toxic standard-setting as resolvable by science" (Wagner 1995, 1661). First, judicial requirements that agency "decisions be based on 'substantial evidence' or at least not be 'arbitrary [and] capricious'" deter agencies from admitting uncertainty, and provide "inescapable incentives to conceal policy choices under the cover of scientific judgments and citations" (Wagner 1995, 1661–1664). Second, courts tend to defer to agency expertise when the issue is framed as scientific in nature.[15]

> By insisting on technical justifications on the one hand, and pledging not to scrutinize the accuracy of the technical explanations on the other, the courts not only fail to prevent the science charade, they make it almost obligatory. (Wagner 1995, 1666)

Additionally, statutes often include deadlines by which agencies must take action. While Congress has little leverage for forcing agencies to meet those deadlines, interested parties can file a lawsuit to force an agency to take action. Courts routinely set deadlines—often as part of settlement negotiations—for issuing standards, and many environmental rules in particular are promulgated subject to judicial deadlines. Unlike statutory deadlines, the executive branch takes court-ordered deadlines very seriously. For better or worse, judicial deadlines force agencies to rearrange their priorities, and focus resources on certain areas at the expense of others. They also serve to limit the amount of analysis, interagency discussion, and public comment on those policies.

NAAQS ILLUSTRATION

The process by which EPA sets National Ambient Air Quality Standards (NAAQS) for "criteria pollutants" under the Clean Air Act illustrates many of the problems identified here. First, Congress directs EPA to set the standards to achieve noble goals, but restricts the agency from considering key factors, establishing instead the pretense that science is sufficient to determine a single point concentration that is "requisite to protect public health." Combined with tight deadlines, the statutory language permits Congress to take credit for laudable public goals, while blaming the executive branch's execution for any undesirable outcomes. The courts have reinforced a limited interpretation of the Act, as well as tight deadlines for issuing revised standards. Executive branch career and policy officials respond by developing scientific-sounding explanations to justify one standard over another. Analysts have an incentive to downplay rather than reveal the implications of key science policy choices, and decision makers point to science as either requiring a new standard or as being so uncertain that a new standard cannot be set. The interagency review process is often truncated by very short timeframes and the limited range of options established by EPA and its Clean Air Science Advisory Committee. Public interveners vigorously defend alternative standards based on their own interpretation of the science.[16]

This has evolved into an adversarial process, characterized by harsh rhetoric in which each party claims the science supports its recommended policy outcome and questions opponents' credibility and motives, rather than a constructive discussion regarding appropriate assumptions and data and the reasonableness of the statutory goal. The real reasons for selecting a non-zero standard are not transparent.

SUMMARY

The institutional environment in which regulations aimed at reducing risks are developed is plagued with perverse incentives, leading to a lack of transparency, misdirected resources, delays, and both over- and under-regulation.

As the BPC's 2009 report, "Improving the Use of Science in Regulatory Policy" observed,

> Political decision-makers should never dictate what scientific studies should conclude, and they should base policy on a thorough review of all relevant research and the provisions of the relevant statutes. But some disputes over the "politicization" of science actually arise over differences about policy choices that science can inform, but not determine. (BPC 2009, 4)

Science plays a fundamental role in efforts to manage risks and protect public health. The interpretation of that science through the science policy process, however, is often misunderstood or even hidden from sight. The science policy process of identifying and choosing scientific studies on which to rely, of characterizing exposure-response relationships, of identifying exposure pathways and alternative risk management approaches needs to be examined from as many angles as possible in the same way other policy decisions are scrutinized. A broad agreement that analytic approaches are appropriately identified, applied, and characterized may reduce the adversarial nature of moving forward. Efforts to identify and characterize the uncertainty in scientific evidence by quantifying the range of outcomes of potential regulatory actions[17] may provide useful data for improving science policy choices and increasing confidence in decisions.

RECOMMENDATIONS FOR REFORM

While people decry the "politicization" of science (Mooney 2006), the problem may really be that we ask too much of science in addressing policy problems (BPC 2009). As the BPC observed:

> A tendency to frame regulatory issues as debates solely about science, regardless of the actual subject in dispute, is at the root of the stalemate and acrimony all too present in the regulatory system today. (BPC 2009, 11)

When questions involving policy judgment and values are characterized as scientific, a small number of people have an effective monopoly on the information that is used and how it is characterized, leading to decisions that are not as accountable or as transparent as they should be.[18] As Coglianese observed, "When regulators purport to rely on science as the sole basis for their policy choices, the real reasons justifying their choices remain hidden from public view" (Coglianese 2009). This is exacerbated by the adversarial nature of rulemaking, and group dynamics that discourage differences of opinion and lead to poor decisions that mask uncertainty and give short shrift to important factors and perspectives.

Successful reforms must address both the conduct of regulatory science and alter the incentives of the parties to the rulemaking process. The BPC focused its recommendations on process reforms "that could help clarify for both officials and the general public which aspects of disputes are truly about scientific results and which concern policy" (BPC 2009, 4). The BPC report observes:

> distinguishing between science and policy is not always easy or straightforward, and scientists may make choices based on values in the course of their work. Nonetheless, policy debate would be clarified and enhanced if a systematic effort were made to distinguish between questions that can be resolved through scientific judgments and those that involve judgments about values and other matters of policy when regulatory issues comprise both. This transparency would both help force values debates into the open and could limit spurious claims about, and attacks on science. (BPC 2009, 15)

We concur with the BPC's identification of the problem,[19] and strongly endorse its recommendations that "new procedures for establishing regulatory policy must . . . clarify which aspects of a regulatory issue are matters of science and which are matters of policy (e.g., economics or ethics)" (BPC 2009, 11). Our recommendations build on the BPC's, recognizing that successful reforms must alter the incentives participants in the regulatory development process have to characterize policy decisions as purely scientific.

We discuss three general types of reforms: 1) those focused on improving the content of regulatory science by strengthening its quality and transparency, 2) those directed at procedural changes that could alter the culture and incentives to blur the lines between science and policy, and 3) those that stimulate the exchange and expansion of knowledge. In each of these general categories, we offer specific recommendations.

IMPROVING THE CONTENT OF REGULATORY SCIENCE

As discussed above, numerous science policy choices influence the numbers that ultimately flow from a risk assessment and into the regulatory arena. Science policy is necessary in the face of alternative, scientifically plausible choices of data, endpoints, and models necessary to estimate risk. When different choices have the potential to have very different implications for risk, which is usually the case, the science policy choices often determine the regulation.

ENHANCING TRANSPARENCY OF SCIENCE AND SCIENCE POLICY CHOICES

A simple change to the application of science policy could quickly increase transparency in distinguishing science from science policy. Instead of making multiple science policy choices in the assessment process, agency scientists should calculate and present multiple risk estimates based on scientifically plausible data sets, endpoints, models, etc. This would be the science. Science policy would come in when one estimate was identified as the agency's preferred choice. The science policy rationale would then be clear to all. This approach would allow all interested parties to see the range of potential risk estimates and understand the risk reduction that might result from a regulatory intervention based on the risk estimate chosen based on science policy grounds and scientifically defensible alternatives.

COMPETING SCIENCE POLICY VIEWS—INTERAGENCY REVIEW

Congress and presidents have established executive branch procedures designed to bring rigor and transparency to the analysis supporting regulations (both the risk assessment and risk management components) and OMB and the regulatory agencies have undertaken several related initiatives. With respect to risk assessment, President Clinton's Executive Order 12866 (E.O. 12866) expresses several principles, including:

> In setting regulatory priorities, each agency shall consider, to the extent reasonable, the degree and nature of the risks posed by various substances or activities within its jurisdiction. (Clinton 1993, Section 1[b][4])

> Each agency shall base its decisions on the best reasonably obtainable scientific, technical, economic, and other information concerning the need for, and consequences of, the intended regulation. (Clinton 1993, Section 1[b][7])

Further, in developing regulations, federal agencies should consider " . . . how the action will reduce risks to public health, safety, or the environment, as well as how the magnitude of the risk addressed by the action relates to other risks within the jurisdiction of the agency" (Clinton 1993, Section 4[c][1][D]).

President Obama's Executive Order 13563 reinforced E.O. 12866 and said that the "U.S. regulatory system . . . must be based on the best available science" (Obama 2011, Section 1[a]). He also retained procedures, in place since 1981, requiring agencies to submit significant rules for review by OIRA, in the Office of Management and Budget, before publication in the *Federal Register* in proposed or final form. The President explicitly recognized OIRA's "checks and balances" function, noting that the office provides a "dispassionate and analytical 'second opinion' on agency actions" (Obama 2009). OIRA enlists other agencies with expertise in specific areas to review not only the draft regulations, but the analytic, scientific and science policy justification on which proposed approaches are based. In this coordinating role, it works closely with other White House offices, particularly the Office of Science and Technology Policy on science matters. Once OIRA has completed its review of a proposed rule, the agency may then publish it in the *Federal Register* for public comment.

Agency regulatory analyses (required for regulations with significant economic impacts of $100 million or more per year) are guided by OMB's Circular A-4, "Regulatory Analysis," which provides analytical tools for issues ranging from understanding and expressing uncertainty, to measuring and reporting quantifiable benefits and costs (U.S. Office of Management and Budget 2003). The *Government-Wide Information Quality Guidelines* provide policy and procedural guidance to federal agencies for ensuring and maximizing the quality of the information they disseminate (U.S. Office of Management and Budget 2002). The *Information Quality Bulletin for Peer Review* provides further guidance for pre-dissemination review of influential scientific information (U.S. OMB 2004). The *Final Bulletin for Agency Good Guidance Practices* (Good Guidance Bulletin) establishes policies and procedures for the development, issuance, and use of significant guidance documents by executive branch departments and agencies and is intended to increase the quality and transparency of agency guidance practices and the significant guidance documents produced through them. *Updated Principles for Risk Analysis* reiterates the risk analysis principles released by OMB in 1995, which addressed risk assessment, risk management, and risk communication, and reinforces them with more recent guidance from the scientific community.

Each of these has probably increased transparency and the quality of scientific and regulatory analyses to a certain extent. Some argue that these procedures have slowed the promulgation of important health and safety regulation (McGarity 1992; Ackerman and Heinzerling 2005), though the evidence of "ossification" is mixed (Johnson 2008). Each of these procedural reforms has had limited success in exposing the science charade, however, probably because they don't change the fundamental incentives for participants in the regulatory development process, as outlined above, as agencies can often follow the letter of the procedures without fundamental change. Also, these requirements are generally subordinate to statutory and judicial constraints placed on agencies' decisions, and judicial deadlines truncate interagency review.

There are two ways to enhance this function. First, make public the views of the science and science policy choices in regulatory science assessments that are developed by OIRA experts and experts from other executive branch agencies. This would not necessarily entail revealing deliberative discussions among experts within the executive branch, which could chill frank debate and constructive exploration. However, making transparent the range of plausible science policy choices (including relevant data and assumptions) could help interested parties understand the strengths, weaknesses and implications of the science policy position advanced in an agency's choice of the "favored" risk estimate from the range of values produced. Second, all regulatory assessments should be accountable for complying with the requirements of Circular A-4, the *Updated Principles for Risk Assessment* and other quality objectives. A clear, publicly available "scorecard" would provide both a guide to the rigor of assessments and an incentive to agencies to meet these generally agreed goals.

PEER REVIEW

Peer review is often considered a fundamental component of the scientific process. In December 2004, OMB issued a memorandum establishing guidelines for the use of external peer-review at all federal agencies and departments (U.S. OMB 2004). It details the reasons that enhanced peer review procedures are necessary for government science documents and sets forth requirements for peer review that each agency and department is expected to implement. In some cases, these requirements were very similar to existing agency practice and for other entities they implied a significant change in procedures. For example, EPA has traditionally made extensive use of external peer review for its assessments (U.S. EPA 2000). Depending on the complexity and stakes of an assessment, the agency may arrange a letter peer-review process, engage the EPA Science Advisory Board (SAB) or, occasionally, request a review by the National Research Council of the National Academy of Sciences.

The OMB Peer Review memo sparked strong reactions both in support[20] and in opposition (e.g., Shapiro, 2004). Contrary to the sort of peer review with which many academic scientists are familiar, however, the OMB requirements did not specify the need for a referee, someone to judge whether an agency or department responded appropriately to the peer review comments. This means that, although a peer review process may occur, there is no discipline in place to ensure that the advice of the scientific community is incorporated in a government risk assessment.

We have four suggestions to strengthen the role of peer review in ensuring the appropriate use of science in analyses underlying regulatory decisions. First, it is critical to recognize that the use of peer-reviewed studies in assessments is only a first step. The key for regulatory use of science is peer review of the analyses themselves including the identification of data, the weighing of competing studies, and appropriateness of the models and measures employed. Recent National Academy of Sciences reviews of EPA assessments has suggested that the Agency needs considerable improvement in this area (National Research Council 2006; National Research Council 2010; National Research Council 2011). Importantly, the charge to a peer review group must ask if the analysis is scientifically and analytically appropriate, not whether the agency followed its science policy prescriptions (e.g., assessment guidelines). In keeping with the suggestion above, science policy choices should be applied after the most scientifically objective and appropriate analysis has been conducted and subject to peer review.

Second, the comments of peer reviewers should be made public and easily accessible. Third, an agency should be required to prepare a response-to-comments document, similar to those prepared in responding to peer review of a scholarly article, that details the way in which peer reviewer comments have been addressed in finalizing the assessment. The response to comments document should also be easily accessible to the public. Finally, the peer reviewers themselves should be given the opportunity to comment on the agency's response-to-comments document and final assessment. These comments would also be available to the public.

IMPROVING THE PROCESS OF REGULATORY SCIENCE—CULTURE AND INCENTIVES

Identifying enhancements to the practices used in applying science to regulatory decisions or increasing the transparency of the process only gets us part of the way. A bigger challenge is addressing the incentives and cultures of participants in the rulemaking process. We put forward several ideas to address incentives but recognize that this is a difficult and long-term problem.

LEGISLATION SHOULD RECOGNIZE THAT DECISIONS INVOLVE POLICY AS WELL AS SCIENTIFIC INFORMATION

These are very complex issues, and it will not be easy for policy debates to recognize the complexity of risk assessment and the fallacy of overly simplistic goals that (1) ask too much of science; and (2) do not recognize tradeoffs when making risk management decisions.

For legislators to make the effort to elevate the debate above simple rhetoric, they must have different incentives and expectations of rewards than exist now. Currently, there is no feedback loop to reward a politician for tackling these issues openly and seriously. Members of the media and academics could help by publicly rewarding political discussion and actions that honestly recognize the uncertainties in risk assessment and tradeoffs involved in difficult ("risk management") decisions about reducing risk and protecting public health, but there is little incentive on their part to do so.

Some statutes directed at health, safety, and environmental risks have facilitated more rational regulatory policy than others by recognizing that risk management requires normative judgments that consider tradeoffs. For example, the Safe Drinking Water Act requires EPA to consider the costs as well as the benefits of requiring local water authorities to install controls for specific substances. Perhaps that is one reason why the debates over drinking water standards are generally less acrimonious than debates over ambient air quality standards. Since the statute allows explicit consideration of tradeoffs when setting standards, the full burden of decision-making is not vested in the risk assessment. As a result, policy makers and interested parties may have less incentive to embed policy preferences in the risk assessment portion of the analysis, because they can debate them openly and transparently in the risk management discussion.

AGENCIES SHOULD STRIVE TO MAXIMIZE THEIR SCIENTIFIC REPUTATION

Confidence in the science underlying regulatory decisions is critical to their success. However, this may be an area of vulnerability for the regulatory process. For example, in a recent survey of members of the Society of Toxicology, more scientists believed EPA overstates the size of risks than those that thought the Agency accurately assessed risks.[21] An important incentive for the civil servants in regulatory agencies must be the protection and enhancement of their scientific reputation and that of the agency. Yet, when agency staff are insulated from others with different experiences, expertise, and views, they are more susceptible to tunnel vision, and likely to discount information and data that do not confirm their prior views. Exposure to perspectives of the broader scientific community, including different scientific disciplines, might make them more aware of different perspectives and approaches, and also more accountable to scientific norms. Implementing our suggestions for improving the practice of peer review would help agencies demonstrate alignment with, and responsiveness to, the general scientific community.

POLICY OFFICIALS MUST BE WILLING TO MAKE POLICY DECISIONS

We identified above the motivations that drive policy officials in regulatory agencies to perpetuate the science charade. Implementation of our suggestion for separating science from science policy, namely developing and making available all scientifically plausible estimates of risk *before* identifying the agency's favored estimate based on a single set of assumptions that embed key science policy choices, would draw back the curtain that allows the science charade to exist. Policy officials will need to choose specific numerical values and publically defend the science policy choices that support a specific choice from a range of scientifically plausible risk estimates. This would provide a serious incentive for policy officials to look into estimates of risk, consult with a broad variety of experts to understand the range of scientific views, and explicitly articulate the policy preferences informing their decisions.

REQUIRE EVALUATION OF REGULATORY IMPACTS

Few regulatory programs are subject to rigorous evaluation and feedback. Most regulatory analyses make predictions about the benefits (in terms of reduced risk) that will accrue from a specific intervention. In the case of environmental, health, and safety regulations these are usually reductions in injury, deaths, and disease. Follow-up to evaluate whether the predicted effects of the regulation were realized would be a powerful incentive to improve the use of science for predicting the benefits of interventions. Of course the same could be done with costs to provide feedback on the accuracy of benefit-cost analyses.

Section 812 of the Clean Air Act Amendments of 1990 required EPA periodically to assess the benefits and costs of the Act (U.S. EPA 2011). Although this is a step in the right direction, it is clear that it does not provide the information necessary to feed back to risk assessment assumptions and procedures. A useful evaluation would measure population changes with respect to the predicted outcome following the regulatory intervention. If reductions in a toxic pollutant were predicted to reduce cancer rates, those rates would be measured. If reductions matched predictions it would provide confirmation that the assessment approaches were reasonable. If the "real life" experience didn't match the predictions, that would provide powerful information to refine risk assessment approaches.

EPA's Office of Research and Development has begun some efforts in this area under the rubric of "accountability science."[22] One of the goals of this effort is ascertainment of actual outcomes, the changes in rates of death or disease in a population, following regulatory action.

A REGULATORY HIPPOCRATIC OATH?

Many have lamented the tunnel vision and lack of consideration of tradeoffs and competing risks inherent in much regulatory decision making (Breyer 1993; Graham and Weiner 1995; Calandrillo 2001). Yet the incentives and the cultures of regulatory agencies encourage a single minded focus on what is seen as the group's mission. This is what leads EPA, for example, deliberately to build a "protective" bias into estimates of risk. This practice arose in the context of making highly focused decisions, often setting standards, but it complicates and distorts efforts to compare risks (Ohanian, Moore, Fowle, Omenn, Lewis, Gray, and North 1997).

What if instead the incentives and culture of agencies were based on a health and safety version of the Hippocratic Oath (first, do no harm)? This ethos would require a careful evaluation of the health and environmental effects of specific regulatory action (or inaction). To weigh competing risks, regulators would want to have accurate assessments, not those with some undefined level of "protectiveness" built in. Interagency review could provide a very valuable function by identifying potential "side effects" of specific regulatory interventions and potentially finding better solutions.

The goal in this case would be a regulatory system that ensured actions would appropriately address the health, safety, and environmental risks at hand while eliminating or minimizing the offsetting risks.

REFORMS THAT STIMULATE THE EXCHANGE AND EXPANSION OF KNOWLEDGE

If a regulatory agency had complete knowledge about the risks of a particular substance or practice, as well as complete knowledge regarding the consequences of strategies to reduce that risk, it might be able to design a comprehensive regulation that perfectly aligned protective measures and hazards, striking just the right balance between the cost and benefit of mitigation. Of course, no one has such information about health and environmental risks, anymore than they do about anything else. Hayek identified the central problem facing public policy as "the unavoidable imperfection of man's knowledge and the consequent need for a process by which knowledge is constantly communicated and acquired" (Hayek 1945, 11).

Hayek's focus was on economic planning and he showed that decentralized markets focus disperse information—information that no one individual can obtain—and convey it efficiently to market participants. Many of the risks of concern to regulatory agencies may not be accounted for in market transactions, however.[23] In these cases, we may require a different solution to Hayek's problem that relevant facts are never possessed by a single mind, to take advantage of knowledge "that is dispersed among many people."[24]

New media may provide a vehicle to address the Hayekian "knowledge problem," stimulating a broader exchange of ideas and expanding our knowledge. Traditional ways of organizing people and information evolved because they were cost-effective at the time (Coase 1937). But as new media platforms reduce transaction costs, they can significantly lower the costs of gathering and aggregating information, and remove obstacles to collaboration across a wide spectrum of individuals (Shirky 2008).

E-rulemaking provides a new platform for following and commenting on federal regulations. Regulations.gov is a government website that makes information available on regulations under development across the federal agencies. It provides a portal for public comments to be filed on proposed rulemakings, and posts those comments for other interested parties to view. In theory, commenters could build on previously posted comments, questioning assumptions, correcting errors, and presenting alternate data and perspectives. Despite this potential, we have not seen much interaction and iteration in public comments, and the explanation may lie both in the Regulations.gov design, and tradition of public comment.

The public generally has a relatively short window between the time it first sees an agency proposal (with its accompanying justification) and the deadline for filing public comment (60 days for most economically significant regulations). The majority of comments on a regulation are submitted on or close to the deadline, which limits the potential for ongoing discussion and iterative collaboration (Balla 2011). Furthermore, commenters often craft their submission with an eye toward future litigation if the final regulation is inconsistent with their perspective. Public comments are legal documents that reflect on the submitter and must be fine tuned before posting.[25]

To address these drawbacks and harness the wisdom of dispersed knowledge, agencies or outside parties[26] might experiment with a collaborative "wiki" approach to public comment, where, rather than each individual or group filing comments in parallel and the agency responding to those comments individually, it could provide a forum for diverse individuals to build on each other's information, adding, editing, updating, and correcting to engage the wisdom of dispersed knowledge on issues where no one person has complete information. Larry Sanger, founder of Wikipedia, calls this "distributed knowledge collaboration." (Sanger 2006)

One big advantage of a wiki approach is what Shirky calls its "publish-then-filter" model, where editing is done after something is posted, rather than before. Participants don't need to worry that their post is incomplete or may have inaccuracies because other participants can expand or correct it.

> In a system where anyone is free to get something started, however badly, a short, uninformative article can be the anchor for the good article that will eventually appear. Its very inadequacy motivates people to improve it; many more people are willing to make a bad article better than are willing to start a good article from scratch. (Shirky 2008, 122)

What sorts of questions are most conducive to a wiki approach? There may be several stages during a rulemaking that would benefit from distributed knowledge collaboration. A wiki process may provide an opportunity to gather input during scientific analyses or advanced notices of proposed rulemaking to help agencies identify available research and information and address the myriad data and science policy questions involved. Engaging public input through a wiki is an intriguing possibility that holds the potential to revolutionize how agencies gather information on which to base public policies.

CONCLUSION

Numerous reputable bodies have recognized that current procedures for developing regulations addressing health and environmental risk blur the lines between science and policy, hindering not only public policy decisions, but development of scientific knowledge itself. Yet, current institutions provide incentives to bury policy judgments in analyses that are presented as science, perpetuating the "science charade" (Wagner 1995). Altering these incentives is challenging. We have offered modest suggestions aimed at 1) improving the content of regulatory science and increasing transparency; 2) improving the process of regulatory science, by strengthening the checks and balances provided by different participants in the rulemaking process; and 3) engaging a broad range of expertise and perspectives to counter the problems insular decision-making brings. More research is needed to identify institutional changes that would motivate participants involved in developing regulatory policy to reveal the uncertainties inherent in risk assessment and clarify "which aspects of disputes are truly about scientific results and which concern policy" (Bipartisan Policy Center 2009, 4).

NOTES

1. "Risk assessors might be faced with several scientifically plausible approaches (e.g., choosing the most reliable dose-response model or extrapolation beyond the range of observable effects) with no definitive basis for distinguishing among them. The earlier Committee [NRC 1983 (The "Red Book")] pointed out that selection of a particular approach under such circumstances involves what it called a *science-policy* choice. Science policy choices are distinct from the policy choices associated with ultimate decision-making. . . . " National Research Council. *Science and Judgment in Risk Assessment*. [S.l.]: Nat'L Academies, 1994. Print. Page 27.

2. Conservatism can be equated with the "protective" approach favored by EPA. For example: National Research Council (1983); National Research Council (1994); National Research Council (2009).

3. For example, the FDA places very little weight on the human relevance of chemically induced tumors in mouse liver. EPA, on the other hand, frequently uses evidence from mouse liver tumors in judging an agent to be a likely human carcinogen and often relies on liver tumors as the endpoint for quantitative risk assessment (Gold, Slone, and Ames 1998, 383).

4. "One of the overarching weaknesses of the draft assessment was a lack of critical analysis of the data on which EPA relied in evaluating methodologic strengths and weaknesses. That lack was particularly evident in the assessment of the epidemiologic data: study selection and conclusions appeared to be based heavily on results that showed positive associations, and other data and the strengths and weaknesses of the selected studies were not adequately taken into consideration. The committee observed similar problems in its review of EPA's evaluation of the genotoxicity evidence, in which preference appeared to be given to studies that reported positive results. Specifically, the EPA did not analyze studies critically with respect to their methodologic strengths and weaknesses, nor did it organize its discussion clearly to provide an integrated consideration of the weight of evidence on the genotoxicity of tetrachloroethylene." National Research Council (2010) Page 3; "While the advantages of the Steenland dataset are described, the Draft [EtO] Assessment contains no list of the criteria that were utilized to select studies for inclusion in the risk assessment process. For example, a description of what constituted adequate sample size, exposure assessment, minimum length of employment, length of follow-up, lag time for selected outcomes, etc., would be helpful." US EPA Science Advisory Board (SAB) Ethylene Oxide Review Panel Draft Advisory Report, 8-30-07, page 30; "The committee was unable to reach consensus on the selection of the critical cancer end point. The majority of the members judged that the uncertainties associated with MCL (particularly the high background incidence, uncertainty about the dose-response relationship, and poor understanding of mode of action) were too great to support using MCL data rather than data on hepatic or renal cancer for determining quantitative estimates of risk. Those members judged that the use of the MCL data could be justified only if it is EPA's policy to choose the most conservative unit risk when considering options but that such justification should be distinguished as a policy decision, not a scientific one. They believed that a more scientifically defensible approach would be to use the dataset that has the least uncertainty rather than the dataset that yields the highest estimate of risk" National Research Council (2010, 18).

5. The questionable fitting of the multistage model to some candidate datasets and insufficient consideration of alternative models contribute to underestimation of the overall uncertainties; EPA does not present the full ranges of variation and uncertainty

in relation to model choice, in large part because it applied only linear or nearly linear dose-response models to all candidate datasets. National Research Council (2010, 9).

6. See Executive Orders 13563 and 12866 governing regulatory analysis and oversight. www.whitehouse.gov/sites/default/files/omb/inforeg/eo12866/eo13563_01182011.pdf

7. The Congressional Review Act of 1996, available at: http://www.archives.gov/federal-register/laws/congressional-review/.

8. According to the Congressional Research Service, 38 percent of the House and 55 percent of the Senate hold law degrees. Eighteen representatives (no senators) have a PhD, and twenty members of the House and four senators have a medical degree (Manning 2011).

9. Section 113 of the Clean Air Act requires EPA to set National Ambient Air Quality Standards (NAAQS), expressed as ambient concentration levels based on levels "requisite to protect" human health and welfare. Clean Air Act of 1990 (Title 42).

10. Legislation not only provides political benefits when it is enacted, but "legislators can reap additional benefits by attacking lax implementation of the laws by administrative agencies." (Percival 1991, 196). "No politician ever lost votes by denouncing the bureaucracy" (Wilson 1989, 235).

11. Wilson refers to this as "marrying the natives" (Wilson 1989, 261).

12. Executive Order 12866 directs agencies to "select those approaches that maximize net benefits . . . unless a statute requires another regulatory approach," and to follow its regulatory principles, "to the extent permitted by law."

13. "We must accept that in government, as in business, people will pursue their own private interest, and they will achieve goals that are reasonably closely related to those of the stockholders or of the citizens only if it is in their private interest to do so. Of course, this penchant does not mean that most people, in addition to pursuing their private interests, have no charitable instincts or tendencies to help others and to engage in various morally correct activities. Yet, the evidence seems strong that these are not motives upon which we can depend for the motivation of long-continued efficient performance" (Tullock 2002). "People are the products, not only of their biology, family and schooling, but of their organizational position. . . . " (Wilson 1989).

14. OMB issued guidelines in 2004 and agencies have adopted or adapted those for their own use. These are discussed more later in this chapter.

15. In *Chevron U.S.A., Inc. v. Natural Resources Defense Council, Inc* ., the Supreme Court established a two-step test for determining whether to grant deference to a government agency's interpretation of a statute. Under the second step, if Congressional intent is not clear, "the issue for the court is whether the agency's answer is based on a permissible construction of the statute." Chevron U.S.A. v. NRDC, 467 U.S. 837 (1984). "Chevron effectively places the agency's thumb on the judicial scales, which should not be tolerated when the interpretation before the court is one that advances the agency's own self interest" (Armstrong 2004).

16. States, which are responsible for implementing NAAQS, are active interveners. Ironically, the states unable to comply with current standards are often more supportive of stricter standards than the states that are in attainment. They have little to lose from stricter standards, but they gain relative to competing states which will have nonattainment conditions imposed on them (*Wall Street Journal* 2008).

17. www.epa.gov/ord/ca/quick-finder/accountability.htm

18. As the BPC report cautions, "The first impulse of those concerned with regulatory policy should not be to claim 'the science made me do it' or to dismiss or discount scientific results, but rather to publicly discuss the policies and values that legitimately affect how science gets applied in decision making" (BPC 2009, 4).

19. "Political decision-makers should never dictate what scientific studies should conclude, and they should base policy on a thorough review of all relevant research and the provisions of the relevant statutes. But some disputes over the "politicization" of science actually arise over differences about policy choices that science can inform, but not determine" (BPC 2009, 4).

20. For example, Ruhl and Salzman specifically argues that because regulatory science *uses* rather than *does* science, peer review is necessary even when analyses rely on previously peer-reviewed scientific evidence (Ruhl and Salzman 2006).

21. www.toxicology.org/pr/AreChemicalsPR.pdf

22. www.epa.gov/ord/ca/quick-finder/accountability.htm

23. Note that E.O. 12866 states that "Federal agencies should promulgate only such regulations as are required by law, are necessary to interpret the law, or are made necessary by compelling public need, such as material failures of private markets to protect or improve the health and safety of the public, the environment, or the well-being of the American people."

24. E.O. 13563 Section (2)(a) states: "Regulations shall be adopted through a process that involves public participation. To that end, regulations shall be based, to the extent feasible and consistent with law, on the open exchange of information and perspectives among State, local, and tribal officials, experts in relevant disciplines, affected stakeholders in the private sector, and the public as a whole."

25. Agencies will sometimes issue an advanced notice of proposed rulemaking (ANPR) to solicit public information on a particular risk without identifying regulatory options. A greater use of ANPRs early in a regulation's development could elicit more wide-ranging scientific information on risks of concern.

26. For example, OpenRegs.com is a privately initiated website established to provide information on regulations open for public comment in a format that is arguably more user-friendly and informative than Regulations.gov.

REFERENCES

Ackerman, Frank, and Lisa Heinzerling (2004). *Priceless: On Knowing the Price of Everything and the Value of Nothing*. New York: The New Press.

Armstrong, Timothy K. (2004). "Chevron deference and agency self-interest." *Cornell Journal of Law and Public Policy* 13(Spring): 203–85.

Balla, Stephen J. (2011). *Public Commenting on Federal Agency Regulations: Research on Current Practices and Recommendations to the Administrative Conference of the United States*. Washington, DC: The Administrative Conference of the United States. www.acus.gov/wp-content/uploads/downloads/2011/04/COR-Balla-Report-Circulated.pdf

Baron, R. S., B. Brunsman, S. L. Hoppe, C. F. Kao, B. Linneweh, and D. Rogers (1996). "Social Corroboration and Opinion Extremity." *Journal of Experimental Social Psychology* 32(6): 537–60.

Bipartisan Policy Center (2009). *Improving the Use of Science in Regulatory Policy* . Washington, DC: Science for Policy Project, Bipartisan Policy Center. www.bipartisanpolicy.org/sites/default/files/BPC%20Science%20Report%20fnl.pdf

Breyer, Stephen (1993). *Breaking the Vicious Circle: Toward Effective Risk Regulation.* Cambridge: London: Harvard University Press.

Brock, W. J., V. L. Dellarco, G. M. Gray, R. W. Lane, J. V. Rodricks, and A. Rulis (2003). "Food Safety: Risk Assessment Methodology and Decision Making Criteria." *International Journal of Toxicology* 22(6): 435–51.

Calandrillo, S. P. (2001). "Responsible Regulation: A Sensible Cost-Benefit, Risk Versus Risk Approach to Federal Health and Safety Regulation." *Boston University Law Review* 81.

Chevron U.S.A., Inc. v. NRDC, 467 U.S. 837 (1984).

Coase, R. H. (1937). "The Nature of the Firm." *Economica*, New Series 4(16) (November 1937), doi : 10.1111/j.1468-0335.1937.tb00002.x, www.jstor.org/stable/2626876

Coglianese, Cary (2009). U.S. Congress. House Committee on Science and Technology. Subcommittee on Investigations and Oversight. *The Role of Science in Regulatory Reform.* April 30, 2009. gop.science.house.gov/Media/hearings/oversight09/april30/coglianese.pdf

Coglianese, Cary (2002). "Empirical Analysis and Administrative Law." *University of Illinois Law Review*. 2002: 1111–1138;

Dudley, S. E. (2010). "Regulatory consultation in the United States." *The George Washington University Regulatory Studies Center*. Washington, DC. www.regulatorystudies.gwu.edu/images/pdf/us_regulatory_consultation.pdf

Editorial (2008). "Beverly Hillbully." *Wall Street Journal*, May 19, 2008. online.wsj.com/article/SB121115921730002453.html?_nocache=1304019635143&mg=com-wsj.

Executive Office of the President, Office of Management and Budget (2007). *Final Bulletin for Agency Good Guidance Practices*. www.whitehouse.gov/sites/default/files/omb/memoranda/fy2007/m07-07.pdf.

Executive Office of the President, Office of Management and Budget (2007). *Updated Principles for Risk Analysis*. www.whitehouse.gov/sites/default/files/omb/memoranda/fy2007/m07-24.pdf.

Executive Order 12866. "Code of Federal Regulations," pt. 8, title 3: 51735–51744.

Executive Order 13563. "Code of Federal Regulations," pt. 8, title 3: 3821–3823.

Federal Advisory Committees Database, (2010). www.fido.gov/facadatabase/default.asp (accessed May 21, 2010).

Felter, Susan, and Michael Dourson (1998). "The Inexact Science of Risk Assessment (and Implications for Risk Management)." *Human and Ecological Risk Assessment* 4, no. 2, 245–52.

Fraas, A. G. (2010). "The Treatment of Uncertainty in EPA's Analysis of Air Pollution Rules." Discussion Paper. *Resources for the Future*. www.rff.org/documents/RFF-DP-10-04.pdf

Gold, L. S., T. H. Slone, and B. N. Ames (1998). "What do Animal Cancer Tests Tell Us about Human Cancer Risk? An Overview of Analyses of the Carcinogenic Potency Database." *Drug Metabolism Reviews* 30(2): 359–404.

Gold, L. S., B. R. Stern, T. H. Slone, J. P. Brown, and N. B. Manley (1997). "Pesticide Residues in Food: Investigation of Disparities in Cancer Risk Estimates." *Cancer Letters* 117: 195–207. potency.berkeley.edu/pdfs/PesticideRisk97.pdf.

Graham, J. D. (1996). "Going Beyond the Single Number: Using Probabilistic Risk Assessment to Improve Risk Management." *Human and Ecological Risk Assessment: An International Journal* 2(4): 1008–1034.

Graham, J. D. and Wiener, J. B., eds. (1995). *Risk vs. Risk: Tradeoffs in Protecting Health and the Environment*. Cambridge: Harvard University Press.

Gray, George M. (1994). "Complete Risk Characterization." *Harvard Center for Risk Analysis: Risk in Perspective* 2(4).

Hayek, F. A. (1945). "The Use of Knowledge in Society." *American Economic Review* 35(4): 519–30.

Hough, Andrew (2010). "Volcanic Ash Cloud Restrictions 'Embarrassing,' Say Airlines." *The Telegraph*, April 19, 2010. www.telegraph.co.uk/travel/travelnews/7605794/Volcanic-ash-cloud-restrictions-embarrassing-say-airlines.html.

Huber, P. (1983). "Exorcists vs. Gatekeepers in Risk Regulation." *Regulation*, November/December, 23–32.

Huber, P. (1983). "The Old-New Division in Risk Regulation." *Virginia Law Review* 69(6): 1025–1107.

Jackson, Lisa (2009). U.S. Congress. Senate. Committee on Environment and Public Works. *Statement of Lisa P. Jackson: Nomination Hearings*. January 14, 2009.

Janis, Irving L. (1972). *Victims of Groupthink: A Psychological Study of Foreign-Policy Decisions and Fiascoes*. Chicago: Houghton Mifflin Harcourt.

Johnson Stephen M. (2008). "Ossification's Demise? An Empirical Analysis of EPA Rulemaking from 2001–2005." *Environmental Law* 38: 767–92.

Lubbers, Jeffrey S. (2006). *A Guide to Federal Agency Rulemaking*. Chicago: American Bar Association.

Manning, Jennifer E. (2011) "Membership of the 112th Congress: A Profile," CRS Report R41647, March 1, 2011, available at: www.senate.gov/reference/resources/pdf/R41647.pdf.

McGarity, T. O. (1992). "Some Thoughts on Deossifying the Rulemaking Process." *Duke Law Journal* 41(6): 1385–1462.
Mooney, C. (2006).*The Republican War on Science*. New York: Basic Books.
Murcott, T. (2009). "Science Journalism, Toppling the Priesthood." *Nature* 459: 1054–1055.
National Research Council (2006). *Health Risks from Dioxin and Related Compounds: Evaluation of the EPA Reassessment*. Washington, DC: National Academies Press.
National Research Council (2010). *Review of the Environmental Protection Agency's Draft IRIS Assessment of Tetrachloroethylene*. Washington, DC: National Academies Press.
National Research Council (2010). *Review of the Environmental Protection Agency's Draft IRIS Assessment of Formaldehyde*. Washington, DC: National Academies Press.
National Research Council and the Committee on the Institutional Means for Assessment of Risks to Public Health (1983). *Risk Assessment in the Federal Government: Managing the Process*. Washington, DC: National Academies Press.
National Research Council, Committee on Risk Assessment of Hazardous Air Pollutants, Board on Environmental Studies and Toxicology, and the Commission on Life Sciences (1994). *Science and Judgment in Risk Assessment*. Washington, DC: National Academies Press.
National Research Council, Committee on Improving Risk Analysis Approaches Used by the U.S. EPA, Board on Environmental Studies and Toxicology, and the Division on Earth and Life Sciences (2009). *Science and Decisions: Advancing Risk Assessment*. Washington, DC: National Academies Press.
Obama, Barack H. (2009). Memorandum on Regulatory Review, January 30, 2009. Federal Register 74 p. 5977, February 3, 2009.
Ohanian, E. V., J. A. Moore, R. J. Fowle III, G. S. Omenn, S. C. Lewis, G. M. Gray, and D. W. North (1997). "Risk Characterization: A Bridge to Informed Decision Making." *Fundamental and Applied Toxicology* 39(2): 81–88.
Percival, Robert V. (1991). "Checks without Balance: Executive Office Oversight of the Environmental Protection Agency." *Law and Contemporary Problems* 54(4): 127–204.
Reilly, John, Peter H. Stone, Chris E. Forest, Mort D. Webster, Henry D. Jacoby, and Ronald G. Prinn (2001). "Uncertainty and climate change assessments." *Science* 293(5529): 430–33.
Ruhl, J. B., and James Salzman (2006). "In Defense of Regulatory Peer Review." *Washington University Law Review*.
Sanger, Larry (2006). "Toward a New Compendium of Knowledge." www.citizendium.org/essay.html.
Schoenbrod, David (1995). *Power Without Responsibility: How Congress Abuses the People Through Delegation*. New Haven: Yale University Press.
Shapiro, Sidney A. (2004). "OMB's Dubious Peer Review Procedures." *Environmental Law Reporter* 34: 10064–10071.
Shirky, Clay (2008). *Here Comes Everybody: The Power of Organizing Without Organizations*. New York: Penguin Press.
Stigler, George (1971). "The Theory of Economic Regulation." *Bell Journal of Economics* 2(1): 3–21.
Sunstein, Cass R. (2009). *Going to Extremes: How Like Minds Unite and Divide*. New York: Oxford University Press.
Tullock, Gordon (2002). *Government Failure: A Primer in Public Choice*. Washington, DC: Cato Institute.
U.S. Food and Drug Administration (2012). *Advisory Committees*. www.fda.gov/AdvisoryCommittees/default.htm.
U.S. Environmental Protection Agency. Office of Science Policy (2000). *Science Policy Council Handbook: Peer Review Handbook*. EPA 100-B-00-001.
U.S. Environmental Protection Agency Charter, Clean Air Scientific Advisory Committee, (2009). yosemite.epa.gov/sab/sabproduct.nsf/WebCASAC/currentcharter?OpenDocument.

U.S. Environmental Protection Agency. Office of Air Quality Planning and Standards (2006). *National Ambient Air Quality Standards for Particulate Matter, Regulatory Impact Analysis: Chapter 5.* www.epa.gov/ttn/ecas/regdata/RIAs/Chapter%205--Benefits.pdf.

U.S. Environmental Protection Agency. Office of Research and Development (2009). "Accountability Research: Assessing the Impact of Air Quality Regulations."

U.S. Environmental Protection Agency. Office of Science Policy (2000). *Science Policy Council Handbook: Risk Characterization.* EPA 100-B-00-002.

U.S. Environmental Protection Agency. Risk Assessment Forum (2005). *Guidelines for Carcinogen Risk Assessment.* EPA/630/P-03/001F.

U.S. Environmental Protection Agency. Risk Assessment Forum (1999). *Guidelines for Developmental Toxicity Risk Assessment.* EPA/600/FR-91/001.

U.S. Environmental Protection Agency. Office of the Science Advisor (2004). *Risk Assessment Principles and Practices* EPA/100/B-04/001.

U.S. Environmental Protection Agency. Office of the Science Advisor (2004). *Staff Paper Risk Assessment Principles & Practices.* Washington, DC: Office of the Science Advisor.

U.S. Environmental Protection Agency (2011). *The Benefits and Costs of the Clean Air Act from 1990 to 2020: Summary Report.* www.epa.gov/oar/sect812/feb11/summaryreport.pdf.

U.S. Environmental Protection Agency (2009). *Using Probabilistic Methods to Enhance the Role of Risk Analysis in Decision-Making with Case Study Examples.* EPA/100/R-09/001.

U.S. Office of Management and Budget (2003). *Circular A-4.* www.whitehouse.gov/sites/default/files/omb/assets/omb/circulars/a004/a-4.pdf.

U.S. Office of Management and Budget (2002). "Government-Wide Information Quality Guidelines." *Federal Register* 67(36). www.whitehouse.gov/sites/default/files/omb/fedreg/reproducible2.pdf.

U.S. Office of Management and Budget (2004). *Information Quality Bulletin for Peer Review.* www.whitehouse.gov/sites/default/files/omb/memoranda/fy2005/m05-03.pdf.

Viscusi, W. Kip (1997). "Alarmist Decisions with Divergent Risk Information." *The Economic Journal* 107(445): 1657–1670.

Wagner, W. E. (1995). "The Science Charade in Toxic Risk Regulation." *Columbia Law Review*, 95: 1613.

Wilson, J. Q. (1989). *Bureaucracy: What Government Agencies Do and Why They Do It.* Basic Books.

EIGHT

A Return to Expertise?

A Proposal for an Institute of Scientific Assessments

Gary E. Marchant

The dominant paradigm of government administration in the early part of the twentieth century was that of neutral, expert government bureaucrats who, if kept walled off from the more political elements of government, could be trusted to make decisions in the public interest. This model of bureaucratic expertise was discredited over time by the legal realist understanding that all governmental decisions are political and subject to individual and organizational interests, biases, and beliefs. These criticisms of the bureaucratic expertise model are surely valid, and are now part of the conventional lore of administrative law and policy today.

It is therefore with some trepidation that I propose a limited return to the bureaucratic expertise model by calling for the creation of a governmental Institute of Scientific Assessments (ISA) to perform science-based assessments of risks and other scientific inputs into government decisions (see also Macbeth and Marchant 2008). My proposed return to scientific expertise is limited in two important respects. First, I limit my proposal to scientific inputs, based on an assertion that science, while never completely value-free, can come closest to achieving the bureaucratic expertise model than other types of more policy- and value-laden governmental decisions. Second, I do not suggest that any governmental decision could be based purely on such scientific inputs but always involves economic, political, ethical, and other value-based inputs (Coglianese and Marchant 2004). Rather, my suggestion is limited to a single input into regulatory decision-making, which is an assessment of what science can and cannot tell us about the issue at hand.

This proposal runs counter to the current trend of integrating the various inputs into decisions into a more holistic whole, as well as the growing belief in the inherently social and political dimensions of science advanced by the social studies of science (SSS) and science, technology, and society (STS) perspectives. For example, Sheila Jasanoff asserts:

> Although pleas for maintaining a strict separation between science and politics continue to run like a leitmotif through the policy literature, the artificiality of this position can no longer be doubted. Studies of scientific advising leave in tatters the notion that it is possible, in practice, to restrict the advisory process to technical issues or that the subjective values of scientists are irrelevant to decision-making. The negotiated and constructed model of scientific knowledge, which closely captures the realities of regulatory science, rules out the possibility of drawing sharp boundaries between facts and values or claims and context. (Jasanoff 1990, 231)

In my view, the trend towards "negotiated and constructed" scientific knowledge has had the effect of encouraging and even legitimizing the political bending and manipulation of scientific input into government decisions. In part at least due to this trend, the reputation of science used in government decision-making has never been lower. This diminution of the role and credibility of science comes at a time of rapidly emerging technological and scientific advances where the role and importance of science in government decision-making has never been greater. There needs to be a more credible way to gather, synthesize, and communicate the scientific inputs into government decision making.

I. IS THERE A PROBLEM?

The government's handling of scientific information has been the subject of growing controversy over the past few decades, with frequent charges of politicizing science and "junk science" against regulators in both Democratic and Republican administrations by political opponents. Given that the science underlying most regulatory issues is incomplete, uncertain, and contested, it is not surprising that an interest group opposed to an administration's regulatory decision would take issue with its scientific conclusions, and make allegations that science has been skewed or politicized. There can be no doubt that regulatory agencies, like attorneys in private lawsuits, interpret and structure uncertain and often conflicting scientific data to support their preferred outcome or position. This process of construing scientific evidence to support a particular position is inherent in the existing system in which the same governmental entity is responsible for the scientific determinations and policy choices.

There are many anecdotal examples of courts or other oversight entities finding that regulatory agencies had adopted a slanted or outcome-driven interpretation of scientific data. For example, in 2000 a reviewing court set aside a U.S. Environmental Protection Agency (EPA) regulation that set a drinking water standard for chloroform based on a linear no-threshold risk model, even though the scientific evidence, as confirmed in two opinions from the EPA's own Scientific Advisory Board, demonstrated that chloroform did not conform to the linear, no threshold dose-response model (*Chlorine Chemistry Council v. EPA* 2000). The Department of Interior's Office of Inspector General found that a senior agency political appointee without any scientific training had "bullied, insulted, and harassed the professional staff of the U.S. Fish and Wildlife Service (FWS) to change documents and alter biological reporting regarding the Endangered Species Program" in favor of a politically preferred outcome (*see* Ruhl 2007). A February 2008 federal court decision recounts how an EPA scientist, David Lewis, who had raised concerns about the potential risks of sludge land application, was harassed and then forced to resign by agency officials who allegedly politicized and manipulated scientific data to support a conflicting position (*McElmurray v. USDA* 2008). Another court decision found that the decision by the Food and Drug Administration (FDA) to deny over-the-counter availability of Plan B emergency contraceptive to seventeen-year-olds, ostensibly based on scientific evidence of risk to such individuals, instead "stemmed from political pressure rather than permissible health and safety concerns" (*Tummino v. Torti*, 2009).

More generally, there is a widespread perception that regulatory agencies do not apply credible science. Public confidence and trust in the government's handling of science is at an all-time low (Bauer 1995). Congress is increasingly skeptical of scientific actions of the agencies they oversee—in the words of one Congressman (Vern Ehlers [R-Mich]) with widely respected expertise on scientific matters, "members of Congress . . . do not have confidence that the agency uses science appropriately in its decisions" (Ehlers 2002, 30). *Nature*, one of the world's leading scientific journals, recently described the U.S. EPA as displaying "a reckless disregard" for science (Nature 2008). According to a former Deputy Administrator of EPA, "[t]he bottom line is that nobody likes EPA science. Congress does not like it, the scientific community does not like it, the environmental groups do not like it, and industry certainly does not like it" (Sussman 2004, 583). Even EPA's own expert committee examining the agency's scientific performance concluded that "EPA science is of uneven quality, and the agency's policies and regulations are frequently perceived as lacking a sound scientific foundation" (EPA 1992). Not to be outdone, the FDA's recent evaluation of its own scientific activities concluded that "science at the FDA is in a precarious position: the Agency suffers from serious scientific deficiencies and is not positioned to meet current or emerging regulatory responsibilities" (FDA 2007, 2). Given the widespread mistrust in agency science, there is a need and opportunity to consider a different approach in how regulatory agencies incorporate science into their decisions.

II. CAN SCIENCE BE OBJECTIVE?

One response to the growing disenchantment with the objectivity of regulatory agency science is to recognize that science is inherently a subjective and politicized endeavor, and thus we should simply accept that science inevitably will be manipulated to support policy outcomes. As a human endeavor, scientific practice and communication is an inherently political, value-laden and biased undertaking. In this regard, the SSS and STS critics of science have a point—the interpretation and application of science is socially constructed. As a discipline, however, science seeks to socially construct its own undertaking to try to minimize human influences and to strive for "objectivity" through core principles identified by early sociologists of science like Robert Merton, such as disinterestedness, skepticism, and data primacy (Merton 1973). While the purity and objectivity of science is an ideal that can never be fully achieved in practice, the critical point for the present discussion is that the context and institutional structure in which science is considered can have a major impact on how close or far a particular scientific determination will be from the scientific ideal.

For example, in the litigation context, scientific testimony presented by court-appointed neutral experts tends to be much more objective and balanced than the science presented by the parties' "hired gun" experts, who are often selected and counseled to provide one-sided absolute opinions that give no credence to the opposing views (Deason 1998). Agency science advisory committees that are composed of individual scientists selected for their expertise and instructed to act as independent scientists regardless of who they are employed by or affiliated with tend to give much more balanced and useful scientific opinions than advisory committees made up of scientists from various interest groups who are selected to represent those constituencies (Ashford 1984). National Academy of Science's committees established and operated on a consensus-seeking model provide much more balanced and objective scientific views than do a small cluster of individual scientists hand-picked by the two political parties to testify at a Congressional hearing. In short, context and institutional structure matter for how closely scientific practice approaches the scientific ideal.

As anyone who has worked with scientists knows, if you gather a group of scientists in a room and let them discuss an issue amongst themselves, even if (and maybe particularly if) they have conflicting positions and views going in, the norms of scientific discourse will tend to dominate, including the primacy of data, the importance of replication, the recognition of uncertainties and limitations of the data, a reluctance to speculate beyond the data, and the need to word any conclusions or recommendations cautiously and with appropriate caveats. In the end, the group will naturally tend to gravitate toward consensus. As described by Bernard Goldstein:

> A scientist advising the Administrator of EPA on a basic risk assessment issue would attempt to use a consensus approach, as would the National Academy of Sciences or a similar body . . . The consensus approach model assumes that the scientists on the panel will learn from each other as they listen to evidence, and weigh and discuss different viewpoints. Interestingly, at the end of this deliberative process, these scientists will usually end up with opinions closer to the center (mean) of this bell-shaped curve than they would if they were simply asked for their individual opinions. After discussing the subject, scientists would move toward a central consensus since most scientists intuitively huddle together on questions of this nature. (Goldstein 1989)

This dynamic is well-illustrated by a written account of the "neutral" scientific panel (consisting of four highly qualified and respected scientists) by Judge Pointer who was overseeing the multi-district litigation (MDL) consolidated cases in the silicone breast implant litigation. The first meeting of the expert panel was conducted under legal rather than scientific rules of conduct:

> The panel's mission began in an unremarkable manner. Lawyers for the women and for the breast-implant makers . . . summarized their arguments before the panel members, who simply listened, while lawyers performed as they would before a jury. (Judson 1999)

In other words, the first meeting was a typical legal hearing, involving scientific evidence, but with the science presented and argued in an adversarial and advocacy context. By the second meeting, things began to change, and the scientific mode began to come into the forefront:

> Judge Pointer announced (with visible relish) that he was hoping the proceedings . . . would resemble a scientific meeting rather than a court of law. Experts selected by both the plaintiffs and defendants were to present their evidence to the members of the science panel as they would do to colleagues at a conference—and the members of the panel would be permitted to ask them whatever questions they liked. Moreover (and the judge's eyes twinkled with glee as he said it), the lawyers would *not* be allowed to interrupt, object, or intervene. They could, however, cross-examine the various experts on their testimony once the members of the science panel had finished asking questions. (Judson 1999)

Eventually the scientific model took over, noticeably affecting how the very same scientists operated and behaved:

> By the following November, when the final hearings were held, the lawyers had completely lost control of the process—something that, as one lawyer remarked, "scared the bejesus out of us." This time, the experts who testified were invited by the scientists on the panel, and the lawyers were not allowed to cross-examine anyone. They simply sat and watched. . . . Observers commented on the experts' change in demeanor during the hearings. In their direct testimony and responses to the panel members' questions, the experts adopted the open presentation style appropriate to a scientific seminar, but they shifted to a grudging bunker-style testimony when the opposing attorneys began the cross-examination. . . . [T]he same scientists who testified fully and openly in responding to questions from their peers on the science panel became evasive and tense when the hostile questioning from the attorneys began. (Judson 1999)

While this is just one example, it makes a more general point that the context and environment in which science is discussed and communicated makes an important difference in how closely that scientific discourse approaches the Mertonian ideal of scientific objectivity (while again recognizing that it will never get there entirely).[1]

III. SEPARATING SCIENCE FROM NON-SCIENCE

It has become accepted wisdom that science cannot be separated from other inputs such as policy, economics, ethics, values, and other factors, all of which overlap and intersect in myriad ways. For example, while risk assessment has traditionally been seen as primarily a scientific undertaking, the current trend is to recognize the important non-scientific policy and value judgments that enter into risk assessment. This issue goes back to an influential 1983 report by the National Research Council (NRC), the research arm of the U.S. National Academy of Sciences, entitled *Risk Assessment and Risk Management in the Federal Government*, known colloquially as the "Red Book" for its red cover, that set forth a framework for regulatory risk analysis that has generally been followed ever since by U.S. and many foreign regulatory agencies (NRC 1983).

A central premise addressed by the "Red Book" was the issue of separating risk assessment, a primarily scientific undertaking, from risk management, a more policy-related undertaking. The Red Book found that "[a]t least some of the controversy surrounding regulatory actions has resulted from a blurring of the distinctions between risk assessment policy and risk management policy," and accordingly recommended that "regulatory agencies take steps to establish and maintain a clear conceptual distinction between assessment of risks and consideration of risk management alternatives" (NRC 1983, 3). Writing in the same timeframe, William Ruckelshaus, two-time administrator of the EPA, likewise called for risk assessment and risk management to "be separated as much as possible within a regulatory agency" (Ruckelshaus 1983, 1027).

While both the NRC report and Administrator Ruckelshaus recommending the separation of risk assessment from risk management *within* a regulatory agency, the NRC report strongly recommended against dividing risk assessment and risk management into separate institutions because of the need for risk assessors and risk managers to communicate with each other. Separating risk assessment and risk management into separate organizations would impede the necessary communication and cooperation between scientists and policymakers.

Thirty years after publication of the "Red Book," the recommendation to separate risk assessment from risk management has become one of the most controversial and contested proposals in the Report. Many scholars and practitioners argue that risk assessment is not solely a scientific undertaking but inevitably involves important policy assumptions (which the Red Book acknowledged), and therefore should be more closely integrated rather than distanced from risk management. In contrast, the recommendation to keep risk assessment and risk management under the same roof in the same agency has rarely been questioned. Indeed, today's conventional wisdom is that it is inappropriate and ineffective to try to separate risk assessment from risk management (Pielke 2007; Goldstein 2003; Walker 2003; Jasanoff 1990).

IV. INSTITUTIONAL PRECEDENTS

Notwithstanding the current trend in favor of integrating rather than separating science from policy and other factors, there are some useful precedents of institutions that have succeeded in focusing only on science and yet achieving high credibility. Two of these success stories are described below (the European Food Safety Authority and the Health Effects Institute), followed by discussion of another institution (Intergovernmental Panel on Climate Change) which was somewhat less successful in separating science from policy.

European Food Safety Authority (*EFSA*): The EFSA was created in 2002 as a response of the European Union and its member nations to a series of food scares in the late 1990s, including the "mad cow" and dioxin in food fiascos. The occurrence and impacts of these incidents were exacerbated by the lack of a European-wide credible agency to oversee food safety. The EFSA was created to serve as "an independent source of scientific advice and communication on risks associated with the food chain" (EFSA undated). EFSA is based on the assumption that risk assessment can be separated from risk management, with EFSA limited to performing the former but not the latter:

> In the European food safety system, risk assessment is done independently from risk management. As the risk assessor, EFSA produces scientific opinions and advice to provide a sound foundation for European policies and legislation and to support the European Commission, European Parliament and EU Member States in taking effective and timely risk management decisions. EFSA's remit covers food and feed safety, nutrition, animal health and welfare, plant protection and plant health. In all these fields, EFSA's most critical commitment is to provide objective and independent science-based advice and clear communication grounded in the most up-to-date scientific information and knowledge. (EFSA undated)

EFSA provides a broad range of scientific assessments relating to food safety to the regulatory bodies of the European Union (EU) (i.e., the EU Commission and the EU Council of Ministers), and issues such assessments in response to specific requests or "questions" from the regulatory agencies. The EFSA is structured and functions in a mode that seeks to maximize adherence to the scientific goals of neutrality and objectivity. It is run by scientists, seeks to prepare scientific reports using open scientific consultation and consensus, and attempts to limit its activities to the scientific aspects of major policy issues. The EFSA has now issued over 500 scientific assessments in its first decade, and is generally considered a credible and successful institution. Indeed, the primary criticisms to date has been that EFSA refuses to consider social, ethical, and policy issues in issuing its scientific assessments when, for example, assessing the risk of genetically modified foods being considered for approval in the European Union. EFSA does not claim that such considerations are inappropriate, but only that they should be factored into the risk management decisions made by the more political branches of the EU, rather than in scientific risk assessments produced by EFSA.

Health Effects Institute (*HEI*): The HEI is a joint EPA-auto industry funded scientific research and analysis nonprofit organization created in 1980 to provide independent research on air pollution issues. The HEI was established as an attempt to create a credible "neutral broker" between the EPA and auto industry on highly controversial air pollution issues, with original funding coming equally from the EPA and auto manufacturers. The HEI's mandate is to provide "high-quality, impartial, and relevant science on the health effects of air pollution" (HEI undated). While its initial charter anticipated that the HEI would only conduct independent research, it has gradually evolved a role in providing neutral scientific assessments of controversial issues. For example, the HEI was called upon to provide an independent scientific assessment of the evidence for the fine particulate matter standard promulgated by EPA in

1997. The HEI re-analysis was highly influential in helping to resolve an intense scientific and political controversy over that standard. As with EFSA, the HEI's commitment in its organizational structure and procedures to providing a neutral, objective scientific assessment has made it a highly regarded and credible organization (Keating 2001; Jasanoff 1990).

Intergovernmental Panel on Climate Change (*IPCC*): The IPCC is an international scientific assessment body created by the United Nations Environment Programme (UNEP) and the World Meteorological Organization (WMO) in 1998. Its purpose is to provide international scientific assessments of climate change caused by human activities, including projections of future climate change, the causes of climate change, and the impacts of climate change on both human and natural systems (Grant 2006). The IPCC operates on a scientific consensus model, with its reports requiring consensus of the participating scientists that "clearly identif[ies] disparate views for which there is significant scientific [. . .] support, together with the relevant arguments" (IPCC 2005). Moreover, the IPCC assessments are intended to be policy relevant, but not policy prescriptive (IAC 2010).

The IPCC has achieved much success and acclaim, culminating in its winning the Nobel Peace Prize (along with Al Gore) in 2007 for producing "an ever-broader informed consensus" about the impact that human activity has had on global climate (Norwegian Nobel Committee 2007). The IPCC reports have generally been highly influential with scientists, the media, and the public at large (Hulme and Mahoney 2010). Much of the scientific credibility of the IPCC has been attributed to its ability to involve the leading climate change scientists from around the world in its process, and a methodology that seeks to incorporate a range of scientific views and opinions into its assessments (IAC 2010).

Notwithstanding these achievements, the IPCC has been "flooded by criticism" in the past few years (Schiermeier 2010). These criticisms primarily relate to the perceived and actual scientific objectivity and neutrality of the IPCC, which has been called into question by a series of statements and actions suggesting that the IPCC has strayed into the policy realm of being an advocate for stronger governmental action on climate change rather than adhering to its intended mission of being an "honest broker" on what the science can and cannot tell us (IAC 2010; Broder 2010). One example is the statement of the Chairman of the IPCC, R. K. Pachauri, in releasing the IPCC's 2007 report, that "I hope this will shock people and governments into taking more serious action" (Crook 2007). The disclosure of emails from the University of East Anglia revealing attempts by some prominent IPCC participants to shut-out opposing perspectives by some scientific critics further suggests an advocacy dimension. Finally, errors and biases detected in the IPCC reports, such as a well-publicized error relating to the longevity of the Himalayan glaciers, that systematically tend to over-state the magnitude or importance of climate change, suggest a "groupthink" mentality within the IPCC in favor of particular policy goals (Economist 2010; IAC 2010; Schiermeier 2010).

These statements and actions straying into the policy realm have severely undercut the IPCC's scientific credibility and objectivity (Broder 2010; Revkin 2009; Kerr 2007). Rather than being seen as an objective assessor of the science, the IPCC is increasingly seen as playing a role as an "issue advocate" (Pielke 2007; Sarewitz 2010). In the words of one former IPCC participant (John R. Christy), "[i]t just feels like the IPCC has gone from being a broker of science to a gatekeeper" (quoted in Revkin 2009). In response to these criticisms and concerns, the United Nations and the IPCC recently commissioned a review of the process and procedures of the IPCC by the InterAcademy Council (IAC), the organization of national academies of science from fifteen jurisdictions including the United States, United Kingdom, Japan, China, and others, which recommended a series of organizational and structural changes to try to restore the scientific integrity and credibility of the IPCC (IAC 2010). The IAC's report provided the following caution: "Straying into advocacy can only hurt IPCC's credibility" (IAC 2010, 5).

How can the problems encountered by the IPCC be prevented? Ross McKitrick (2012) and Jason Scott Johnston (2012), in their chapters in this collection, argue that the IPCC structure and function is fatally flawed, and any objective assessment is impossible given the inherent conflicts and politicization of scientists participating in the process. My argument is less pessimistic—scientific advice can be structured to maximize its objectivity and utility, although of course, never reaching the perfect objectivity envisioned by Robert Merton. The answer to the problems encountered by IPCC and other scientifc assessment bodies in the past and future is to fix it rather than abolish it, by putting in place institutional autonomy and safeguards intended to maximize the scientific ethos within the institution. In the next section, I begin to lay out some steps that can be taken in this regard.

V. THE PROPOSED INSTITUTE FOR SCIENTIFIC ASSESSMENTS

The examples briefly summarized above suggest that an independent, stand-alone scientific assessment body can provide highly valuable and credible scientific input into the regulatory process if it is structurally and procedurally designed to stick to scientific matters and resist any temptation to stray into policy advocacy. Such a body, if designed correctly, therefore has the potential to greatly enhance the integrity and credibility of the scientific input on which regulatory decisions can be based. An Institute for Scientific Assessments (ISA), as previously proposed by Macbeth and Marchant (2008), could fill such a role. The ISA would be a governmental scientific body, staffed and managed by full-time federal employee scientists, overseen by an external advisory board that would include prominent national scientific experts, such as the leaders of the National Academy of Sciences (NAS) and the American Association for the Advancement of Science (AAAS).

> As Macbeth and Marchant first described its proposed operation and function:
> This new scientific assessment agency would not conduct its own research, but rather would gather, evaluate and assess the existing data in a manner that could be used by a regulatory agency in making decisions. The regulatory agencies could identify questions on which they needed scientific assessments through an annual regulatory agenda, supplemented with ad hoc requests as they arise throughout the year (similar to the EU Commission's requests to EFSA or EPA's occasional requests for scientific reviews by the National Research Council or its own Science Advisory Board). In addition to requesting risk assessments for specific rulemakings, an agency may also request a scientific analysis from the ISA on a more general or cross-cutting issue. Congress could also request a scientific opinion from the ISA, helping to fill the gap in Congressional science advice since the demise of the Office of Technology Assessment in 1995. (Macbeth and Marchant 2008, 162–63)

In conducting its assessment, the ISA would be directed to follow the norms of scientific inquiry as closely as possible, including objectivity, disclosure of uncertainties and competing hypotheses, and consensus-seeking. As with EFSA, the ISA would be prohibited from straying from its scientific mandate into risk management or policy terrains, and enforcing this critical boundary would be one of the primary directives of the ISA management and external advisory board.[2] Again as has been the experience with both EFSA and HEI (and where the IPCC encountered problems), instilling a culture of scientific objectivity from top to bottom of the organization will be critical to the ISA's success.

The ISA approach would provide several benefits to the current dismal state and credibility of government regulatory science. *First,* the ISA structure, limiting consideration to scientific data and issue only, would squeeze out much of the perceived or actual political and policy influence currently afflicting regulatory agency science. While it would be naïve to suggest that the ISA can eliminate altogether the perception or actuality of political influence or bias (Ashford 1988), the examples described above show how the cultural environment in which scientific advice is developed and delivered can have a major impact on how closely the Mertonian ideals of objective science can be approached. *Second,* the ISA approach could reduce the "science charade" (Wagner 1995), which is the tendency of agencies to often try to disguise their policy decisions as scientific determinations in order to protect those decisions from public,

Congressional and judicial scrutiny (Coglianese and Marchant 2004; Rushefsy 1984). Because the ISA would provide a credible independent assessment about what the science does and cannot tell us, it will be much harder for regulatory agencies to camouflage their policy preferences as science. Thus, regulatory decision-making will be more transparent.

A *third* potential benefit of the ISA would be to harmonize scientific assessments of the same issue between different federal agencies. Currently, there are frequently inconsistencies in how different agencies address the same scientific questions (Sadowitz and Graham 1995). These inconsistencies confuse the public and create unfairness and uncertainty as former EPA Administrator William Ruckelshaus described:

> The public interest is not served by two federal agencies taking diametrically opposed positions on the health risks of a toxic substance and then arguing about it in the press. We should be able to coordinate our risk assessment procedures across all federal agencies. The risk management strategies that flow from that assessment may indeed differ, depending on each agency's statutory mandate or the judgment of the ultimate decisionmaker. (Ruckelshaus 1983, 1028)

The ISA could help promote a more uniform treatment. For example, several federal agencies are currently evaluating the toxicity of bisphenol-A (BPA), including the EPA, FDA, and National Institute of Environmental Health Sciences (NIEHS). It is quite possible we could end up with three or more disparate scientific assessments of the risks of BPA by different federal agencies. With the ISA, each of the agencies would utilize the same uniform assessment from the ISA in their regulatory decisions.

Another example where the ISA would be particularly valuable at this time is nanotechnology. Nanotechnology has become a major technological innovation, with over 1000 products already on the market, and many more in development. Several different federal agencies are currently evaluating risks from various nanomaterials. Rather than duplicating efforts and perhaps arriving at inconsistent conclusions, the ISA could provide an integrated scientific assessment of what is known about nanotechnology toxicity that would help to harmonize agency responses (Marchant and White 2011).

The creation of an ISA would no doubt raise a number of administrative and procedural issues. For example, what if a regulatory agency wanted to depart from the scientific findings of the ISA? What opportunity would there be for public comment and perhaps even judicial review of ISA assessments? Could a party challenging in court an agency regulation that relied on an ISA assessment raise claims against the ISA assessment on the merits? These and other issues would require careful consideration. One potential model to emulate however might be the ambient air quality standards promulgated by EPA. The first step in this process is for EPA scientists to develop a "Criteria Document," sometimes referred to as an "Integrated Scientific Assessment," that reviews and summarizes the key findings from the available scientific literature. This Criteria Document is subject to a public comment period, but is not directly reviewable by courts. The Criteria Document, both in terms of its content and procedural status, could be a good model for an ISA scientific assessment. The air quality standard adopted by EPA based on the Criteria Document is reviewable by courts, and a petitioner could conceivably raise arguments about the substance or procedure of the Criteria Document. Such challenges attacking the Criteria Document are rare, as likely would be legal challenges against ISA assessments used to support regulatory decisions.

CONCLUSION

Science is likely to continue to play an ever-increasing role in regulatory decision-making for the foreseeable future. At the same time, there is a growing consensus that the current approach for integrating science into regulatory decision-making is broken. A government-wide scientific authority that can provide independent scientific assessments, such as the Institute for Scientific Assessments proposed here, that have the best chance of approaching the scientific ideal of objectivity and neutrality may be the best option to rebuild credibility in government science.

NOTES

1. In his contribution to this book, Ross McKitrick sets up the same dichotomy between advocacy-based and consensus-based scientific advisory structures, reaches the same conclusion as herein that the current balance between these two approaches is sub-optimal, but then reaches the polar opposite conclusion of this chapter in recommending a greater emphasis on advocacy rather than consenus approaches (McKitrick, 2012). Ultimately, resolving which direction is best will need to be decided by empirical results, which argues in favor of a variety of institutional experiments to test the different approaches.

2. Additional institutional, process, and structural suggestions for enhancing the objectivity for agency risk assessments described by Conrad (2012) and by Dudley and Gray (2012) in this book would be applicable or adaptable to the ISA proposed herein.

REFERENCES

Ashford, Nicholas A. (1988). "Science and Values in the Regulatory Process." *Statistical Science* 3(3): 377–83.

Ashford, Nicholas A. (1984). "Advisory Committees in OSHA and EPA: Their Use in Regulatory Decisionmaking." *Science, Technology, and Human Values* 9(1): 72–82.

Bauer, M. (1995). "Resistance to New Technology and Its Effects on Nuclear Power, Information Technology and Biotechnology," in *Resistance to New Technology*, edited by M. Bauer, 1–43. Cambridge, UK: Cambridge University Press.

Broder, J. M. (2010). "Scientists Take Steps to Defend Climate Work." *New York Times*, March 3, A1, A11.
Chlorine Chemistry Council v. EPA, 206 F.3d 1286 (D.C. Cir. 2000).
Coglianese, Cary, and Gary E. Marchant (2004). "Shifting Sands: The Limits of Science in Setting Risk Standards." *Pennsylvania Law Review* 152(4): 1255–1360.
Crook, C. (2007). "The Steamrollers of Climate Science." *Financial Times*, August 2, 7.
Deason, Ellen R. (1998). "Court-Appointed Expert Witnesses: Scientific Positivism Meets Bias and Difference." *Oregon Law Review* 77(1):59–156.
Editorial (2010). "Science Behind Closed Doors." *Economist*, July 10, 76-77.
Ehlers, V. (2002). "Both Sides are Right: EPA Needs to Improve Science Function." *Environmental Forum*, January/February, 30–31.
Environmental Protection Agency (EPA) (1992). "Safeguarding the Future: Credible Science, Credible Decisions, Report of the Expert Panel on the Role of Science at EPA." Washington, DC: Government Printing Office.
European Food Safety Authority (EFSA) (Undated). EFSA, About EFSA, available at www.efsa.europa.eu/EFSA/efsa_locale-1178620753812_AboutEfsa.htm.
Food and Drug Administration (FDA) (2007). "FDA Science and Mission at Risk," available at www.fda.gov/ohrms/dockets/ac/07/briefing/2007-4329b_02_01_FDA%20Report%20on%20Science%20and%20Technology.pdf.
Goldstein, Bernard D. (2003). "Risk Characterization and the Red Book." *Human & Ecological Risk Assessment* 9(5): 1283–1289.
Goldstein, Bernard D. (1989). "Risk Assessment and the Interface between Science and Law." *Columbia Journal of Environmental Law* 14(2):343–55.
Grant, J. (2006). *A Guide to the Intergovernmental Panel on Climate Change* 4th ed. International Petroleum Industry Environmental Conservation Association, London.
Health Effects Institute (HEI) (undated). "About HEI," available at www.healtheffects.org/about.htm.
Hulme, Mike, and Martin Mahoney (2010). "Climate Change: What Do We Know About the IPCC?" *Progress in Physical Geography* 34(5): 705–18.
InterAcademy Council (IAC) (2010). *Climate Change Assessments: Review of the Processes and Procedures of the IPCC*. IAC, Amsterdam.
Jasanoff, Sheila (1990). *The Fifth Branch: Science Advisers as Policymakers*. Cambridge, MA: Harvard University Press.
Judson, Olivia (1999). "Slide-Rule Justice." *National Journal*, October 9, 2882–84.
Keating, Terry J. (2001). "Lessons from the Recent History of the Health Effects Institute," *Science, Technology & Human Values* 26(4): 409–30.
Kerr, Richard A. (2007). "Pushing the Scary Side of Global Warming." *Science* 316(5830): 1412–1414.
Macbeth, Angus, and Gary Marchant (2008). "Improving the Government's Environmental Science." *New York University Environmental Law Journal* 17(1): 134–69.
Marchant, Gary E., and Andrew White (2011). "An International Nanoscience Advisory Board to Improve and Harmonize Nanotechnology Oversight." *Journal of Nanoparticle Research* 13(4): 1489–1498.
McElmurray v. USDA, 535 F.Supp.2d 1318 (S.D. Ga. 2008).
Merton, Robert K. (1973). *The Sociology of Science: Theoretical and Empirical Investigations*. Chicago: Chicago University Press.
National Research Council (NRC) (1983). *Risk Assessment and Risk Management in the Federal Government*.Washington, DC: National Academy Press.
Nature (editorial) (2008). "The EPA's Tailspin." *Nature* 452(7183): 2.
Norwegian Nobel Committee (2007). "The Nobel Peace Prize for 2007." nobelprize.org/nobel_prizes/peace/laureates/2007/press.html.
Pielke, Jr., Roger (2007). *The Honest Broker: Making Sense of Science in Policy and Politics*. Cambridge, UK: Cambridge University Press.
Revkin, Andrew C. (2009). "Nobel Halo Fades Fast for Climate Change Panel." *New York Times*, August 4, D1.

Ruckelshaus, William D. (1983). "Science, Risk, and Public Policy." *Science* 221(4615): 1026–1028.
Ruhl, J. B. (2007). "Reconstructing the Wall of Virtue: Maxims for the Co-Evolution of Environmental Law and Environmental Science." *Environmental Law* 37(4): 1063–1082.
Rushefsky, Mark E. (1984). "The Nisuse of Science in Governmental Decisionmaking." *Science, Technology & Human Values* 9(3): 47–59.
Sadowitz, March, and John D. Graham (1995). "A Survey of Residual Cancer Risks Permitted by Health, Safety and Environmental Policy." *Risk: Health, Safety & Environment* 6(1): 17–35.
Sarewitz, Daniel (2010). "Curing Climate Backlash." *Nature* 464(7285): 28.
Schiermeier, Quirin (2010). "IPCC Flooded by Criticism." *Nature* 463(7281): 596–97.
Sussman, Robert M. (2004). "Science and EPA Decision-Making." *Journal of Law and Policy* 12(2): 573–87.
Tummino v. Torti, 603 F.Supp.2d 519 (E.D.N.Y., 2009).
Wagner, Wendy E. (1995). "The Science Charade in Toxic Risk Regulation." *Columbia Law Review* 95(7): 1613–1723.
Walker Jr., Bailus (2003). "Impacts of the Red Book." *Human & Ecological Risk Assessment* 9(5): 1373–1382.

Index

About the Contributors

ABOUT THE EDITOR

Jason Scott Johnston is the Henry L. and Grace Doherty Charitable Foundation Professor of Law and the Nicholas E. Chimicles Research Professor in Business Law and Regulation, University of Virginia School of Law. He is the author of recent articles on climate change law and regulation in *Regulation, Notre Dame Law Review,* and the *Harvard Environmental Law Review,* and has published dozens of articles on law economics both in law reviews such as the *Yale Law Journal* and peer-edited journals such as the *Journal of Law, Economics and Organization.* Johnston is coordinator of the Searle Workshops on Markets and the Environment at the University of Virginia School of Law.

ABOUT THE CONTRIBUTORS

James W. Conrad Jr. is a principal of Conrad Law & Policy Counsel. Before forming Conrad Law & Policy in 2007, Conrad was Assistant General Counsel at the American Chemistry Council. Prior to that, he was in private practice with the Washington, D.C., offices of Davis, Graham & Stubbs and Cleary, Gottlieb, Steen & Hamilton, where his responsibilities encompassed regulatory advocacy, counseling, litigation, and transactional work under all the major federal environmental laws. He has published dozens of articles and book chapters on issues in environmental and administrative law in journals such as the *Administrative Law Review* and *Law and Contemporary Problems.*

Susan E. Dudley is a research professor for the Trachtenberg School of Public Policy at George Washington University. She is founder of the George Washington University Regulatory Studies Center. In September 2010, Dudley was appointed as a member of the Administrative Conference of the United States. From 2006 until 2008, Dudley served as administrator of the Office of Information and Regulatory Affairs in the White House Office of Management and Budget. From 2003 until 2007, she was director of the Regulatory Studies Program at George Mason University.

George M. Gray is professor of environmental and occupational health and director of the Center for Risk Science and Public Health at the George Washington University School of Public Health and Health Sciences. From 2005 to 2009 he served as the assistant administrator for the Office of Research and Development and as the science advisor at the U.S. Environmental Protection Agency. Prior to joining EPA, Gray was executive director of the Harvard Center for Risk Analysis and a member of the faculty of the Harvard School of Public Health. Gray's primary research interests are risk characterization, risk communication, and the role of science in policy-making. He has published on both the scientific bases of human health risk assessment and its application to risk policy with a focus on risk/risk tradeoffs in risk management. His professional service includes serving as a councilor for the Society for Risk Analysis, and on the Risk Assessment Task Force, Congressional Task Force and Communications Committee of the Society of Toxicology. Other service includes the NIEHS National Advisory Environmental Health Science Council, and on an IOM Committee evaluating the components of the Women, Infants and Children (WIC) food packages.

Gary E. Marchant is Lincoln Professor of Emerging Technologies, Law, and Ethics and is executive director and faculty fellow at the Center for Law, Science & Innovation in the Sandra Day O'Connor College of Law at Arizona State University. Prior to joining the ASU law faculty in 1999, Professor Marchant was a partner at the Washington, D.C., office of Kirkland & Ellis, where his practice focused on environmental and administrative law. He has published four books and dozens of book chapters and articles (in journals such as the *University of Pennsylvania Law Review*) dealing with issues concerning environmental law and regulation, with a focus on regulatory science.

Ross McKitrick is professor of economics at the University of Guelph in Ontario, Canada. Author of *Taken By Storm: The Troubled Science, Policy and Politics of Global Warming* (2003), he has also published economics articles in journals such as *Environmental and Resource Economics*. His physical science research has appeared in journals such as *Journal of Geophysical Research, Geophysical Research Letters, Climate Research, The Journal of Non-Equilibrium Thermodynamics*, and *Proceedings of the National Academy of Sciences*.

Rob Roy Ramey II is a principal of Wildlife Science International. Ramey received his PhD from Cornell University and is a consulting biologist specializing in issues arising under biodiversity laws and treaties such as the U.S. and California Endangered Species Act. He is the author of dozens of scientific papers published in peer-edited journals such as *Science, Animal Conservation, Molecular Ecology*, and the *Wildlife Society Bulletin*.

Katrina Miriam Wyman is professor of law at the New York University School of Law. A legal scholar specializing in natural resource law, with a particular focus on endangered species regulation, her publications have appeared in leading law journals, such as the *New York University Law Review,* and the book *Breaking the Logjam: Environmental Protection That Will Work* (2010; with David Schoenbrod and Richard Stewart).

www.ingramcontent.com/pod-product-compliance
Ingram Content Group UK Ltd.
Pitfield, Milton Keynes, MK11 3LW, UK
UKHW020125290926
15471UKWH00003B/294